HOLLEY

CARBURETORS, MANIFOLDS & FUEL INJECTION

HOW TO SELECT, INSTALL, TUNE, REPAIR AND MODIFY HOLLEY FUEL SYSTEM COMPONENTS FOR STREET AND RACING USE

MIKE URICH & BILL FISHER

HPBooks

HPBOOKS
Published by the Penguin Group
Penguin Group (USA) Inc.
375 Hudson Street, New York, New York 10014, USA
Penguin Group (Canada), 90 Eglinton Avenue East, Suite 700, Toronto, Ontario M4P 2Y3, Canada
(a division of Pearson Penguin Canada Inc.)
Penguin Books Ltd., 80 Strand, London WC2R 0RL, England
Penguin Group Ireland, 25 St. Stephen's Green, Dublin 2, Ireland (a division of Penguin Books Ltd.)
Penguin Group (Australia), 250 Camberwell Road, Camberwell, Victoria 3124, Australia
(a division of Pearson Australia Group Pty. Ltd.)
Penguin Books India Pvt. Ltd., 11 Community Centre, Panchsheel Park, New Delhi—110 017, India
Penguin Group (NZ), 67 Apollo Drive, Rosedale, North Shore 0632, New Zealand
(a division of Pearson New Zealand Ltd.)
Penguin Books (South Africa) (Pty.) Ltd., 24 Sturdee Avenue, Rosebank, Johannesburg 2196,
South Africa

Penguin Books Ltd., Registered Offices: 80 Strand, London WC2R 0RL, England

While the author has made every effort to provide accurate telephone numbers and Internet addresses at the time of publication, neither the publisher nor the author assumes any responsibility for errors, or for changes that occur after publication. Further, publisher does not have any control over and does not assume any responsibility for author or third-party websites or their content.

Copyright © 1994, 1987 by HPBooks
Drawings by Erwin Acuntius and Holley Carburetors
Photos by Bill Fisher, Mike Urich, Howard Fisher and others
Cover design by Bird Studios
Front cover photo of Holley Pro-Jection throttle body by Ed Monaghan, courtesy of Holley
Replacement Parts Division of Coltec Industries Inc.

The cooperation of Holley Replacement Parts Divsion of Coltec Industries Inc. is gratefully
acknowledged. However, this publication is a wholly independent production of HPBooks.

PRINTING HISTORY
HPBooks trade paperback edition / January 1987
Revised HPBooks edition / June 1994

Library of Congress Cataloging-in-Publication Data

Urich, Mike.
        Holley carburetors, manifolds & fuel injection: How to select, install,
tune, repair, and modify two- & four-barrel carburetors, manifolds and Pro-Jection
fuel injection for street and racing use /
Mike Urich & Bill Fisher.—Rev. and updated.
        p.   cm.
    Rev. ed. of Holley carburetors & manifolds. Rev. ed. c 1987.
    Includes index.
    ISBN 978-1-55788-052-9
    1. Holley carburetors. 2. Automobiles—Fuel systems.   I. Fisher, Bill, 1926–   .
II. Urich, Mike. Holley carburetors & manifolds.   III. Title.
IV. Title: Holley carburetors, manifolds & fuel injection.
TL212.U74   1994   94-49484
629.25'33—dc20   CIP

PRINTED IN THE UNITED STATES OF AMERICA

25th Printing

NOTICE: The information in this book is true and complete to the best of our knowledge. All recommendations on parts and procedures are made without any guarantees on the part of the author or the publisher. Tampering with, altering, modifying or removing any emissions-control device is a violation of federal law. Author and publisher disclaim all liability incurred in connection with the use of this information.

Contents

Introduction

WHY A BOOK ON HOLLEY FUEL INJECTION, CARBURETORS & MANIFOLDS?

First—When the original edition was printed in 1972, no such book was available. Automotive enthusiasts need accurate and tested information on using Holley high-performance carburetors in one volume.

Automotive textbooks have carburetion sections. And, entire books are devoted to carburetor design. But neither type of book really helps the enthusiast because there's not enough practical information with the theory and mathematics.

While there's some carburetor information in auto repair manuals, it is always oriented toward repairing a *normal* car—not one being tuned for *high-performance*, racing or economy. **Second**—Holley Engineering is constantly developing a lot of good information. But to use it, you need a basic understanding of the various systems within the carburetor and how they are interrelated. And, you need more information on their Pro-Jection fuel-injection systems, too.

So, it was a case of Holley not being able to explain important details unless the listener or reader knew how a carburetor or fuel-injection works. Details without the necessary underlying knowledge and understanding can be misleading. They can even cause problems for the user unless he understands the systems relationships—which are not all that obvious.

Holley's low-cost performance team for the small-block Chevrolet is the 0-1850 600 cfm Model 4160 carburetor on the low-profile 300-38 dual-plane intake manifold.

Holley's Pro-Jection 4 four-barrel 900 cfm fuel-injection throttle body mounted on the high-rise 300-36 dual-plane intake manifold.

George Holley in a replica of his 1897 three-wheeler. He built it at age 19!

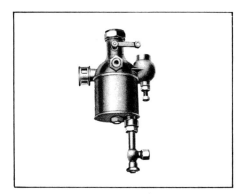

"Iron Pot" carburetor made by Holley Brothers for 1904 curved-dash Oldsmobile started their specialization into carburetor manufacturing. Early customers also included Buick, Pierce-Arrow and Winton.

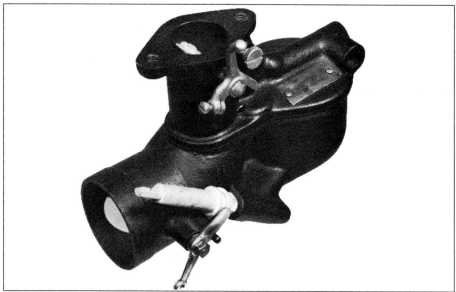

Holley's Model 390 for Model A Ford. Several million were made. Because Zenith was also a supplier, Holley carburetors had an "H" or "Holley" cast in small letters inside bowl. Zeniths had "Z" on outside of bowl.

Third—Holleys are the most widely used high-performance carburetors and fuel-injection systems (aftermarket) in the world. So it was important to create a book describing the important ones in detail and telling how to get the most out of them.

Finally—The past 30 years or so have seen a continuously increasing emphasis on emission controls. Because the fuel system is one of the main controls, understanding its relationship to other systems and components in the automobile is very essential. Everyone needs to know more about how the various systems work—and the role played by the carburetor and fuel injection in meeting emission standards.

Photos and illustrations have been used copiously to describe construction features and operation. Some show how to use standard or special parts to get improved performance.

Factory service and overhaul manuals are seldom available, so we've included how-to photo sequences on installing a manifold and carburetor, disassembly and assembly of Holley high-performance carburetors and fuel-injection system details.

You will find a lot of tips on high-performance carburetors and fuel-injec-tion systems. We incorporated answers to the questions enthusiasts ask again and again at technical seminars and at the racetracks. We have dispelled a great many rumors and myths and half-truths which are part of the romance of using Holley equipment.

Because Holley continually improves their, high-performance and street-legal carburetors, as well as their fuel-injection systems, there will always be changes. There is no way to capture more than a snapshot of development and parts availability at that final moment when the printing presses roll.

So, to keep track of what is happening in high-performance carbure-tion and fuel injection, stay tuned to the availability of new parts and pieces from Holley. Make sure you always have a copy of the latest *Holley High-Performance Parts Catalog.*

HOLLEY HISTORY

ORIGIN

From its very beginnings, Holley Carburetor Company combined an intense interest in racing with a dedi-cation to engineering excellence.

Founded in 1902 by the Holley Brothers of Bradford, Pennsylvania, the company grew out of their experi-ments with the infant horseless carriage.

At the age of 19, George M. Holley designed and built his first car—a single-cylinder three-wheeler capable of 30 miles per hour.

Fascination with speed and things mechanical soon led young Holley into motorcycle racing. He made a name for himself in national competi-tion. Together with his brother Earl, they formed Holley Brothers to build motorcycle engines when they were not racing. Later, they built complete motorcycles.

This unique combination of talents led to still another vehicle—long since disappeared from the auto scene—the Holley *Motorette.* Introduced in 1903. this jaunty little red sports model sold for $550—fully equipped. More than 600 of the 5.5-hp vehicles were built over three years. Only three survive—one is in Holley's lobby in Warren, Michigan.

As the fledgling auto industry was taking shape, the first hint of industrial specialization began to emerge. Sensing this trend, the Holley brothers concentrated on designing and building carburetors and ignition components. They left building basic vehicles to their customers and became original-equipment suppliers to Pierce-Arrow, Winton, Buick and Ford.

FIRST CARBURETOR

Their first original carburetor, called the *iron pot*, appeared on the curved-dash Oldsmobile in 1904.

Over the years, Holley carburetors were found on AMC, Chrysler, Ford, General Motors, International Harvester, Mack, Diamond Reo and other vehicles. And, the carburetors were often used as original equipment on high-performance cars such as the Corvette LT-1, Camaro Z-28 and Mustang. Holley ignition distributors were standard equipment on thousands of vehicles built by International Harvester, Ford and other makers.

$100 per horsepower! "Holley Motorette" Runabout cost $550 with 5-1/2 hp engine. This 64-inch wheelbase car weighed 600 pounds. Features included sight-feed lubrication, planetary-drive transmission and tilting steering wheel with lock. Carburetion? Single-barrel Holley, of course. 600 were made in 1903-1904.

COMPLETE INDUCTION SYSTEMS

Holley makes hundreds of different carburetors and fuel-injection units for replacement applications. Should you want more fuel flow or decide to alter your basic induction system, they probably produce the parts to do it.

Holley carburetors have long been the front runners in the performance area: The original three-barrel, the NASCAR 4500, and a family of large two-barrels, four-barrel double-pumpers, the Model 4165/4175 small/large spread-bore for good emissions with performance, the newer 2010, 4010 and 4011, and the first stand-alone replacement fuel-injection system.

In 1976, a line of street and strip manifolds joined the Holley Induction Team. Holley makes the entire system: fuel pumps, fuel lines, fuel filters, carburetors, air cleaners, intake manifolds and electronic fuel-injection systems and components.

Other items in the Holley line include high-performance electric fuel pumps, ignition kits, valve covers and electrical components. The 1980s saw the transition from carburetors to fuel injection on new cars. Because they had been heavily involved in fuel-injection research since the late '60s, Holley was able to begin supplying original-equipment fuel-injection-system components for new-car manufacturers and for aftermarket fuel-injection rebuilders.

As of 1988, Holley began offering new replacement bolt-on throttle-body injectors that provide exciting performance gains over original equipment.

In 1990 Holley began offering stand-alone throttle-body fuel-injection systems in both two- and four-barrel models. These replace carburetors to provide the benefits of fuel injection for anyone who needs performance and simplified tuning capabilities. In 1992 Holley added a line of closed-loop fuel-injection systems.

As you can see, Holley has always been closely identified with high performance. It was the cornerstone on which the company was built.

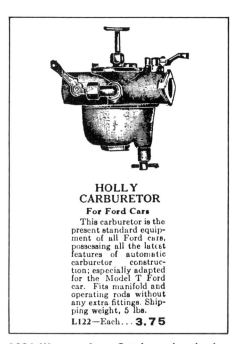

HOLLY CARBURETOR
For Ford Cars
This carburetor is the present standard equipment of all Ford cars, possessing all the latest features of automatic carburetor construction; especially adapted for the Model T Ford car. Fits manifold and operating rods without any extra fittings. Shipping weight, 5 lbs.
L122—Each... **3.75**

1926 Western Auto Catalog advertised a new Holley at a low price! They're worth more than that today. Carburetor is Model NH for Model T Ford.

Performance continues as an important part of Holley Replacement Parts Division of Coltec Industries, Inc.

Engine Requirements

Determine your engine's fuel and air requirements before selecting a carb/manifold combination. Double-pumper 850 cfm 4150 0-8162 on 300-25 Competition manifold is a typical drag-race setup.

This chapter concentrates on explaining some of the variables that affect the air/fuel (A/F) requirements of an engine. The objective is to understand these requirements so you can then apply a carburetor or fuel-injection system that satisfies them. We'll refer to the carburetor or throttle body as a *fuel system, system or unit*.

Don't worry about following complicated formulas because we've distilled fuel-system theory into some fundamental concepts. Let's leave the complex calculations to the Holley design engineers.

AIRFLOW REQUIREMENTS

Because the air an engine consumes has to come in through a carburetor or throttle body, knowing how much air the engine *can* effectively consume will help you select the correct system size.

How big should the system be? Two variables plugged into a simple formula can help determine the correct system size for an engine.

Engine Displacement—specified in cubic inches (cid).

Remember, 1000 cubic centimeters (cc) = 1 liter = 61 cid. Or divide engine displacements specified in cc by 16.4 to convert metric displacement values to cid. For example a 2-liter engine equals 122 cid (2 X 61 = 122) or 2000cc divided by 16.4 equals 122 cid.

Maximum rpm—The peak rpm that the engine will achieve. Be realistic with this value. An inflated figure will cause you to buy too large a carburetor, which will cause problems discussed throughout this book. This is

not nearly as critical with fuel injection, as explained in that chapter, page 191.

Let's assume your engine has perfect "breathing" or 100% *volumetric efficiency* (VE) and apply our values to calculate the airflow requirement for the engine in cubic feet per minute (cfm).

For 2-Cycle Engines

$$\frac{\text{cid} \times \text{rpm} \times \text{Volumetric Efficiency}}{1728} = \text{cfm}$$

For 4-Cycle Engines

$$\frac{\text{cid}}{2} \times \frac{\text{rpm}}{1728} \times \text{Volumetric Efficiency} = \text{cfm}$$

Example: 350 cubic inch engine

8000 rpm maximum
Assume volumetric efficiency of 1 (100%)

$$\frac{350 \text{ CID}}{2} \times \frac{8000 \text{ rpm}}{1728} \times 1 = 811 \text{ cfm}$$

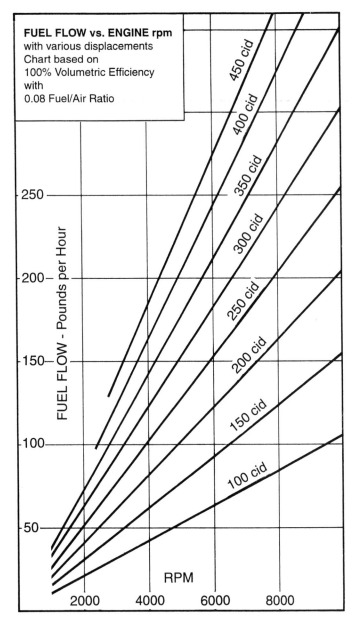

FUEL FLOW vs. ENGINE rpm
with various displacements
Chart based on
100% Volumetric Efficiency
with
0.08 Fuel/Air Ratio

FUEL FLOW - Pounds per Hour

450 cid
400 cid
350 cid
300 cid
250 cid
200 cid
150 cid
100 cid

250
200
150
100
50

RPM
2000   4000   6000   8000

To estimate fuel flow, multiply rpm by VE you expect from engine. Chart is based on full-power F/A ratio of 0.08 (AF = 12.5:1), that is suitable for nearly all engines.

AIRFLOW vs. ENGINE rpm
with various displacements
Chart based on
100% Volumetric
Efficiency

AIRFLOW - cubic feet per minute (cfm)

450 cid
400 cid
350 cid
300 cid
250 cid
200 cid
150 cid
100 cid

1000
800
600
400
200

RPM
2000   4000   6000   8000

To find the engine's airflow requirement, select maximum rpm, cid and note cfm at left. Multiply this value by the VE you expect from the engine (i.e., 0.80). Use a carburetor or injection throttle body with airflow rating equal to or slightly smaller than this airflow requirement.

FLOW RATINGS
   Carburetor or injection throttle-body airflow rating in cfm is a more accurate measurement of flow capacity than comparing venturi sizes. Venturi size does not represent the actual flow capacity. One or more booster venturis may reduce the effective opening and increase the restriction or pressure drop across the carburetor. While the injector throttle body does not use main or booster venturis, the flow can be reduced by any item which reduces airflow, such as the fuel injector and its supports.

A VE of 100% or 1.00 is usually not attainable with a naturally aspirated (unsupercharged) engine. There are losses in the induction system and air pumping capability of the engine that must be accounted for. Thus, our example engine won't flow 810 cfm of air (at standard temperature and pressure). Let's talk about volumetric efficiency and how it affects airflow requirement.

VOLUMETRIC EFFICIENCY (VE)

This value, signified by the Greek letter *eta* (η), is a measure of how well an engine breathes. The better the breathing ability, the higher the VE. VE is really an incorrect description of what is measured. But, the term's usage is established, so it's futile to try to change it to the correct term, *mass efficiency*.

VE is the ratio of the *actual* mass (weight) of air taken into the engine, to the mass the engine displacement would *theoretically* consume *if there were no losses*. The ratio is expressed as a percentage.

$$VE = \frac{\text{Actual mass of air taken in}}{\text{Theoretical mass of air that could be taken in}}$$

VE is quite low at idle and low speeds because the engine is being throttled. It reaches a maximum at a speed close to the point where maximum torque at wide-open throttle (WOT) occurs, then falls off as engine speed is increased to peak rpm. A VE curve closely follows the torque curve of an engine.

Effect of VE–After calculating the airflow requirement for an engine with 100% VE, you'll reduce the cfm value according to the actual VE you expect out of the engine. What percentage VE can you expect?

An ordinary low-performance engine has a VE of about 75% at maximum speed; about 80% at maximum torque. A high-performance engine has a VE of about 80% at maximum speed; about 85% at maximum torque. An all-out racing engine has a VE of about 90% at maximum speed; about 95% at maximum torque.

A highly tuned intake and exhaust system with efficient cylinder-head porting, and a camshaft ground to take full advantage of the engine's other equipment, can provide such complete cylinder filling that a VE of 100% (or slightly higher) is obtained

at the speed for which the system is tuned.

Our example 350-CID engine was calculated as requiring 810 cfm of air at 100% VE. If this is a high-performance engine with a maximum of 85% VE, then the actual airflow requirement becomes 810 cfm x 0.85 = 688 cfm (at standard temperature and pressure).

AIR MASS & DENSITY

Because the mass of air taken in is directly related to air density, VE (η) can be expressed as a ratio of the density achieved in the cylinder (γcyl) to the inlet air density (γcyl inlet), or

$$\eta = \gamma cyl / \gamma i$$

Ideal mass flow for a 4-cycle engine is calculated by multiplying:

$$\frac{rpm}{2} \times CID \times \gamma i$$

where γi is the inlet air density.

Air density varies directly with pressure. The lower the pressure, the less dense the air. At altitudes above sea level, pressure drops, reducing power because the density is reduced.

Tables that relate air density to pressure (corrected barometer) and temperature are available. Or, you can use this formula:

$$\gamma = \frac{1.326P}{t + 459.6}$$

where:

γ = density in lb/cubic feet

P = absolute pressure in inches of mercury (Hg) read directly off of barometer (corrected)

t = temperature in degrees Fahrenheit at induction-system inlet

Actual mass flow into an engine can be measured with a laminar-flow unit or other gas-measuring device as the engine is running. A calibrated orifice or a pitot tube can also be used.

This actual mass flow is usually lower than ideal in a naturally aspirated engine because air becomes less dense as it is heated in the intake manifold. Absolute pressure (and density) also drop as the mixture travels from the fuel-system inlet into the combustion chamber. This further reduces the mass of the charge reaching the cylinder.

The greater the pressure drop across or through the carburetor or throttle body, the lower the density can be inside of the manifold and in the combustion chamber. If the fuel-system unit is too small, pressure drop at WOT will be greater than the desired 1.0 in.Hg for high performance. Power will be reduced because the mixture won't be as dense as it needs to be for full power.

Carburetor or injection throttle-body capacity is one way to state equivalent

MORE AIR, MORE HP

If you're unfamiliar with thinking about the density of the combustion charge as it enters during the intake stroke, all this talk of mass flow, pressure drop, and VE probably becomes confusing. Here is an engine model to think about which can make these concepts clearer.

The engine is an air pump–it ingests air and compresses it. It's often assumed the reason air enters cylinders is because the pistons draw it in–they create a vacuum and that pulls in the air. But, atmospheric pressure (about 15 psi) will push air into a void, provided its passage is non-restrictive.

When a carburetor or fuel-injection throttle body is large and its throttle is wide open, atmospheric pressure helps fill the lower-pressure area above a piston on the intake stroke. Close the throttle to a small opening and atmospheric pressure forces a partial charge of air past the restriction. The result is less-dense air in the consuming cylinder—so, less power.

Consequently, a small venturi and throttle blade opening restrict the amount of air atmospheric pressure can push into the cylinder. For most street and highway driving a small-venturi carburetor is most efficient and gives the best low- and mid-range performance. If your goal is maximum power, and low- and mid-range economy, driveability and torque are secondary, then think big.

Think big all the way. Bigger intake ports, higher lift cam with longer duration, bigger intake valves and an efficient exhaust system to discharge burned mixture properly. Restrictions anywhere in the induction path keep atmospheric pressure from pushing in a full air charge and the engine won't perform to its full potential.

The more air atmospheric pressure can force into the combustion chamber, the more HP.

On this test stand airflow is measured by critical flow orifices; fuel flow is measured electronically. Here, airflow is being measured on Holley's Model 3739 throttle-body injection unit. Test stand is used for carburetors and injection throttle bodies.

size. It is the quantity of airflow through the unit (at standard temperature and pressure) at a given pressure drop. It is 1.5 in.Hg for four-barrel carburetors and all injection throttle bodies and 3.0 in.Hg for one- and two-barrel carburetors. The higher the flow rating,

the bigger the unit. Or, the bigger the unit, the lower the pressure drop across it at any given airflow.

To keep VE as high as possible, we would like to use a large carburetor or throttle body, thereby maintaining a lower pressure drop. *The size limitations are at the other end of the flow curve.* Will the carburetor meter fuel correctly at low airflows? And, will it work OK in the engine's frequently used speed range? This is not a problem with fuel injection because it sprays fuel into the manifold under system pressure and does not rely on pressure drop to meter fuel into the engine.

FUEL REQUIREMENTS

Fuel requirements relate to the airflow requirement because fuel is consumed in proportion to the air used by the engine. Fuel flow is stated in pounds per hour (lb. per hr.) and sometimes in pounds per hp hour (lb. per hp hr.), termed *specific fuel consumption* because it states how much fuel is used for *each* horsepower in one hour.

The relationship between the amount of fuel and the amount of air which flow together into an engine is

called the *fuel/air (F/A) ratio*. This is pounds of fuel divided by pounds of air. An engine uses a lot more air than fuel, so fuel/air ratios are always small numbers such as 0.08, which equals 1 pound of fuel divided by 12 pounds of air.

Some people find this easier to understand if the ratio (0.08) is turned upside down to become an air/fuel *(A/F) ratio*. This gives a number like 12.5:1, meaning the engine uses 12.5 pounds of air for each pound of fuel (or 1.0 divided by 0.08). Obviously, calculations can be made either way because these ratios are just two different ways to state the same unit.

First, let's look at the WOT full-power fuel requirement for our example 350-cid engine that needed 810 cfm of air at 8000 rpm with 100% VE.

To convert airflow into lb. per hr:

cfm x 4.38 = Airflow lb./hr.

Where 4.38 is a factor for 60F at one-atmosphere pressure (14.7 psi).

Multiply by the F/A ratio, which we will assume to be a typical full-

EQUIVALENT RATIO TABLE

A/F (Air/Fuel)	F/A (Air/Fuel)	A/F (Air/Fuel)	F/A (Air/Fuel)
22:1	0.0455	13:1	0.0769
21:1	0.0476	12:1	0.0833
20:1	0.0500	11:1	0.0909
19:1	0.0526	10:1	0.1000
18:1	0.0556	9:1	0.1111
17:1	0.0588	8:1	0.1250
16:1	0.0625	7:1	0.1429
15:1	0.0667	6:1	0.1667
14:1	0.0714	5:1	0.2000

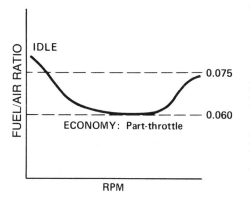

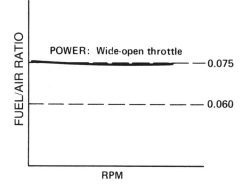

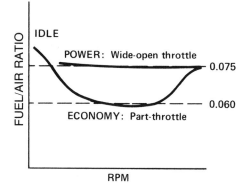

power ratio of 0.077 F/A or A/F ratio of 13:1.

cfm x 4.38 x F/A = Fuel Flow lb./hr.

or

811 x 4.38 x 0.077 = 273 lb./hr.
at 8000 rpm

This is the maximum fuel the engine could consume with 810 cfm airflow. This fuel flow will rarely be reached because—like airflow—fuel flow must be reduced by VE; assume 0.85 VE in this example:

Assuming η = 0.85

cfm x 4.38 x F/A x η = Fuel Flow lb./hr.

or

811 x 4.38 x 0.077 x 0.85 = 232 lb./hr.
at 8000 rpm

Fuel flow is less when the engine is running slower. For instance, 232 lb. per hr at 8000 rpm drops to 1/2 or 116 lb. per hr at 4000 rpm and 58 lb. per hr at 2000 rpm. These consumption measurements assume WOT and the same F/A ratio in each instance.

The chart on page 8 shows fuel flow for various engine sizes over typical rpm ranges. If this book isn't handy when you need to estimate fuel requirements for fuel pump and fuel line selection, here is a rule of thumb:

WOT typically requires 0.5 lb. of fuel per hp every hour. Thus, a 300-hp engine needs 300 X 0.5 = 150 lb. per hr., or 25 gallons per hr. because gasoline weighs 6 pounds per gallon (150 lb. per hr. divided by 6 lb. per gallon equals 25 gallons per hr.).

These calculations are for the maximum fuel consumption rate of an engine under full-power, WOT conditions. This knowledge will help you select the fuel-line sizes, fuel-pump capacity and tank capacity required for your driving application.

Stoichiometric Mixture—This is the ideal fuel mixture—proportioned so all of the fuel burns with all of the air. The exhaust has only carbon monoxide (CO), water vapor (H_2O) and nitrogen (N_2).

Fuel + Air → CO_2 + H_2O + N_2

Stoichiometric mixture is achieved at an F/A ratio of approximately 0.068, or an A/F ratio of 14.7:1.

The actual ratio at which this occurs with an ideal set of conditions varies with the fuel's molecular structure. Gasolines vary somewhat in structure, but not significantly. Fuels other than gasoline require different ratios for the stoichiometric condition.

Methyl alcohol (wood alcohol) has a lower heat content (calorific value) than gasoline and requires a 0.16 F/A ratio (6.4:1 A/F) for its ideal burning condition. This is so much more fuel volume than required for gasoline that most carburetors can't be used with alcohol, except when highly modified. Passages in the carburetor are actually too small to allow correct fuel flow and metering. See the chapter on alcohol modifications, page 137, for more details.

The following comments on fuel requirements for power, economy, idle and cold operation are intended for open-loop conditions, whether a carburetor or fuel injection is used. The carburetor uses mechanical devices and circuits within itself to effect these changes while electronic fuel injection uses output signals to the injectors in response to input signals from various sensors.

Closed-loop or *feedback* systems, whether carbureted or fuel-injected, are a different story. The object in these cases is to keep the mixture or fuel/air ratio at or near the stoichiometric level in order to allow the three-way catalyst to function correctly. There's more detail on these systems in Chapter 2, *How The Carburetor Works*.

Maximum Power—Maximum power requires excess fuel to make sure all of the oxygen in the air is consumed. The reasons for more fuel: mixture distribution to the various cylinders and fuel/air mixing are seldom perfect. Imperfect combustion leads to the formation of oxides of nitrogen (NO_X) and carbon monoxide (CO).

When all of the air enters into the combustion process, more heat is generated and heat means pressure the engine can convert to work. A typical combustion reaction looks like this:

$$Fuel + Air \rightarrow CO_2 + H_2O + CO + HC + N_2$$

where:

CO_2 = carbon dioxide

H_2O = water

CO = carbon monoxide

HC = unburned hydrocarbons (gasoline)

N_2 = nitrogen

The fuel excess usually amounts to 10%–15%, giving F/A ratios of 0.075–0.080 (13.3:1–12.5:1 A/F ratios). Sometimes, an excess of fuel beyond that for producing maximum power is used for internal cooling of the engine. This is done to reduce or prevent knock and detonation. But from a pollution standpoint, unburned HC and CO are undesirable byproducts of this excess.

Maximum Economy—Maximum economy requires excess air to ensure that all of the fuel is consumed. A typical F/A ratio is 0.06 (A/F = 16:1). The combustion reaction looks like this:

$$Fuel + Air \rightarrow CO_2 + H_2O + N_2 \\ + \text{small amounts of} \\ CO + HC + O_2$$

Under heavy loads, any mixture leaner than stoichiometric burns with sufficient heat to cause any free oxygen to combine with the remaining nitrogen to produce oxides of nitrogen (NO_X). This is another undesirable emission product.

At high engine speeds and low loads, mixtures as lean as 0.055 F/A (A/F = 18:1) are sometimes approached when seeking peak economy. This mixture is on the borderline where it begins to become unstable in its ability to burn in normal engines.

Idle—At idle, the problem of exhaust dilution appears. Intake charge dilution is caused by the high manifold

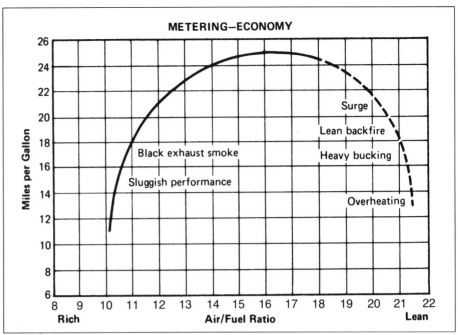

Best fuel economy is achieved at A/F ratio of about 16:1.

Thermocouples in headers (arrows) measure exhaust temperatures for each cylinder to assess distribution characteristics of manifold/carburetor combination Holley engineers strive to keep all EGTs within 100F (38C) spread. Manifold shown here has fabricated fiberglass runners as part of a development test.

12

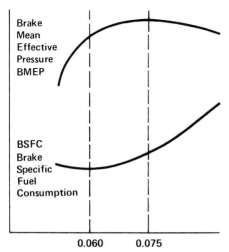

Brake mean effective pressure (BMEP) of engine operating with perfect distribution, showing relationship of F/A ratio supplied by carburetor. For maximum power, 0.075–0.080 F/A yields lowest possible fuel consumption for that output. For maximum economy, 0.060 F/A gives lowest possible fuel consumption consistent with a lower, best-economy output.

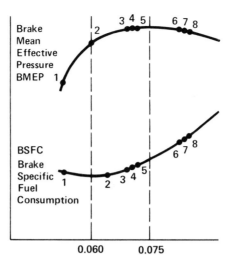

Same curve as at left with numbers representing cylinders receiving various F/A ratios at maximum power output with non-perfect distribution. Cylinders 6, 7, 8 with too-rich F/A produce less power and consume excess fuel; 3, 4, 5 receive correct F/A for best power and consume least fuel for that output; 1, 2 have lean F/A ratio. Richer mixture would have to be supplied to ALL cylinders to bring 1 & 2 to the flat part of curve to avoid detonation. Net result: Only cylinders 1 & 2 will operate at peak power with minimum fuel consumption for that output.

vacuum and by opening the intake valve before the piston reaches top center. When the intake valve opens, some exhaust gases are forced into the intake manifold by the pressure difference, assisted by the still-upward-rising piston.

The exhaust gases dilute and effectively lean the charge. Also, some exhaust gas stays in the cylinder clearance volume. When the piston descends on the intake stroke, the initial charge consists largely of exhaust gas, with a small portion of fresh fuel/air mixture. As the piston completes its intake stroke, the percentage of fresh fuel/air mixture is usually somewhat higher and this portion of the inlet charge burns well enough to supply power for idling the engine.

In the dilution process, some fuel molecules combine (or line up, as the chemists say) with the exhaust mole-

cules and some fuel molecules line up with oxygen in the air. To make sure the mixture is combustible, the mixture has to be made 10%–20% richer than stoichiometric to offset that part of the fuel which combines with the exhaust gas. The richer mixture also helps to offset distribution problems. It also creates additional emission problems because rich idle mixtures generate large amounts of CO.

This is true regardless of whether the fuel is provided by a carburetor or fuel-injection system.

Cold Starting—Starting a cold engine requires the richest mixture of all. Slow cranking speeds create air velocity too low for much fuel vaporization to occur. And, because both the fuel and the manifold are cold, fuel vaporizes poorly.

Vaporization is necessary for combustion, so a lot of excess fuel is

required for starting. Typical F/A ratios are 1.0–2.0. The F/A ratio being supplied isn't what actually gets to the cylinders. Distribution problems and liquid fuel depositing on the manifold walls and on the cylinder-head ports all rob fuel from the mixture. The mixture actually reaching the cylinders in a vaporized form is probably in the 0.06–0.10 F/A range (16:1–10:1 A/F).

Once the engine fires, rpm goes up and velocity through the carburetor increases. There is better vaporization and liquid fuel on the manifold walls vaporizes. As this happens, the mixture gradually leans out to 0.12–0.14 F/A (8:1–7:1 A/F). When the engine warms up, normal operating mixtures can be used.

This relates to carburetion. With fuel injection, fuel is introduced under pressure with fair atomization. Cold-start and cold-running mixtures are significantly leaner.

UNIFORM DISTRIBUTION

We have looked at the engine's fuel and air requirements. Now let's consider another important engine requirement: Uniform (or equal) fuel and air mixture distribution to the cylinders. Uniform distribution helps every phase of engine operation. Non-uniform (unequal) distribution causes a loss of power and efficiency, and may increase emissions.

Uniform distribution relies on a number of factors: Good atomization of the fuel by the carburetor, use of fuel with the correct volatility for seasonal temperature variations, using appropriate techniques to ensure vaporization of the fuel, and correct manifold design. Manifold design is detailed in a separate chapter. We will discuss the other items after we have established the importance of uniform distribution to the cylinders.

Uniform Distribution at Idle—As discussed in the preceding section,

Pontiac 4-cylinder, 2.5-liter "Iron Duke" engine on dynamometer. Exhaust-mani fold probes measure temperature of exhaust gasses as they exit port. EGTs are used to determine cylinder-to-cylinder distribution of fuel/air mixture.

idle F/A is typically 10%–20% richer than ideal to offset dilution by exhaust gas. Idle mixtures with less than 1% CO (a common emission setting) require that all cylinders receive approximately the same idle mixture.

If distribution is not good, getting a reasonably smooth idle may require idle mixture settings that are too rich. This can give CO emissions too high to meet emission regulations.

Uniform Distribution for Economy–Maximum economy, as previously discussed, requires a F/A ratio of approximately 0.06 (16:1 A/F). If one cylinder receives a lean charge due to a mixture-distribution problem, this can increase the amount of fuel required for a given power output. To avoid misfiring in the lean cylinder, the other cylinders must be operated at mixtures richer than the desired 0.06 F/A to bring the lean one up to the 0.06 figure.

All cylinders should be operating at nearly the same F/A because lean cylinders produce more NO_x if they are still firing. And, if they are so lean that misfiring occurs, unburned HC is emitted. Rich-running cylinders pro-

duce excess CO and unburned HC. In either case, undesirable effects include decreased economy and increased emissions.

Uniform Distribution for Power– F/A ratios of approximately 0.075 (13.3:1 A/F) are needed to produce maximum power. If one cylinder runs lean, an excessively rich mixture must be supplied to the other cylinders to bring the lean cylinder up to the correct ratio so it won't run into detonation.

Rich cylinders will waste a lot of fuel and fuel consumption will be increased to maintain a specified power level. Running some cylinders rich reduces their power output. All cylinders must have the same F/A to get the best possible power.

Unequal distribution not only affects fuel consumption and power, it also means ignition timing can only be a rough approximation or compromise of what could be used if all cylinders received equal F/A mixtures. Usable spark advance is directly related to the F/A available in the cylinder.

Timing advance is typically limit- ed so it won't produce knock in the

leanest cylinder. This limits the power that can be developed from those cylinders that have been richened to compensate for unequal distribution.

Check for Correct Distribution–Automobile engine designers (and carburetion designers too) use several methods to check whether the various cylinders in an engine are receiving a uniform mixture. All methods require using a chassis or engine dynamometer. Generally speaking, these methods aren't used by individual tuners because of the cost and complexity of the equipment. In many instances combinations of methods are used.

Distribution is checked by:

• Chemically analyzing exhaust gas samples (combustion products) from the individual cylinders at various operating modes: Idle, cruise, acceleration, maximum power and maximum economy.

There is a definite relationship between F/A and the chemical components in the exhaust gas. By using the accurate analytical equipment developed for emission studies, exact determinations of F/A ratio are obtained to check distribution. The availability of such equipment has made chemical analysis the preferred method automotive engineers use for distribution studies.

• Measuring exhaust gas temperature (EGT) with thermocouples inserted into the exhaust manifold or header at each cylinder's exhaust port. The designer looks for EGT peaks as various main-jet sizes are tried.

If all cylinders were brought to the same temperature, one (or more) cylinder(s) might be 200F (93C) below its peak (and therefore below its peak output). This occurs because all cylinders don't produce the same EGT due to differences in cooling, valves, porting and ring sealing. Perfect distribution would place all cylinders at their peak EGTs with the same jet. This ideal is rarely achieved.

• Studying specific fuel consumption to determine how closely the F/A delivered by a particular carburetor and manifold matches ideal F/A ratios at various operating conditions. Any great variance is cause for suspecting a distribution problem.

• Observing combustion temperatures at various operating conditions.

WHAT AFFECTS DISTRIBUTION?

Atomization & Vaporization— Before gasoline can be burned, it must be vaporized. Vaporization changes the liquid to a gas state and this change only occurs when the liquid absorbs enough heat to boil. For example, a tea kettle changes water to water vapor (steam) by transferring heat into the water until the boiling temperature is reached. At this point, additional heat must be added to change the water into steam. Steam enters the atmosphere as water vapor (a gas, really). This extra heat is

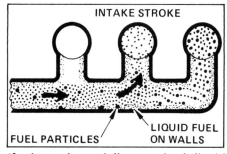

If mixture is partially vaporized, liquid particles clinging to manifold walls (or avoiding sharp turns into cylinder), may cause some cylinders to run lean and some rich. Here center cylinder tends to run lean and end cylinder tends to receive a rich mixture. When possble, it's best to divide mixture before a change in direction occurs. Liquid particles are relatively heavier than the rest of the mixture and tend to continue in one direction. A fully vaporized mixture promotes good distribution in every instance.

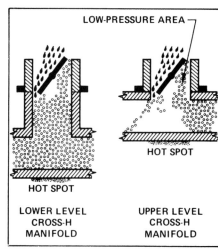

Cross section through V8 engine manifold of Cross-H or dual-plane, two-level type shows how riser height affects distribution. Throttle in this position causes worst distribution.

called the *heat of vaporization.*

Pressure controls the boiling point of any liquid. In the case of water, 212F (100C) is the sea-level boiling point. At higher altitudes, due to lowered atmospheric pressure, water boils at lower temperatures. Remember this relationship. It is important in understanding what happens in the carburetor and manifold.

In most passenger-car engines, heat is applied to the intake manifold to raise the temperature of the incoming mixture.

The higher the temperature, the better the vaporization. There is some sacrifice in top-end power when the mixture is heated sufficiently to ensure vaporizing most of the fuel.

This power loss is because charge density is reduced when the mixture is heated to this point. In most passenger-car engines, the loss is more than offset by the smoother running gained at part throttle. No manifold heat is used in a racing engine because a cold, dense mixture produces more power.

Not only is it necessary to vaporize fuel before it can be expected to burn, it is essential to vaporize it to aid dis-

tribution. It is much easier to distribute vaporized fuel in a fuel/air mixture than it is to distribute liquid fuel. Some liquid fuel is nearly always present on the manifold surfaces. The only time manifold surfaces are "dry" is during high-manifold-vacuum conditions that promote vaporization.

The carburetor discharges gasoline into the air stream as a spray atomized (torn or sheared into fine droplets) into a mist. At this point it should become vaporized. Pressure in the intake manifold will be much lower than atmospheric (except at WOT) and this considerably lowers the boiling point of the gasoline. At the reduced pressure, some fuel particles vaporize as they absorb heat from the surrounding air. And, some fuel particles vaporize when they contact or pass close by the manifold hot spot.

Imperfect Vaporization—This may occur if the mixture velocity is too low, if the manifold or incoming air is cold, if manifold vacuum is low (higher pressure), and if fuel volatility is too low for the ambient temperature. Vaporization is also affected by manifold design and the carburetor size (flow capacity). With the same engine speed, a large carburetor has less velocity through its venturi (air passage). So, there is less pressure drop and a greater tendency for the fuel to come out of the discharge nozzle in liquid blobs or large drops that are not easily vaporized.

Fuel injection doesn't present as many cylinder-to-cylinder distribution problems as does carburetion, but we can't assume that there are no problems. All new fuel system, manifold and engine combinations must be tested for distribution.

The throttle-body injection system's advantage over carburetion, especially at low airflows, is that fuel is delivered at a fairly high pressure drop and there is less chance for the fuel to come out of atomization and puddle. However, we must still be concerned with the directional effect of the throttle plates. Good manifold design is still essential. A uniform conical spray pattern from the fuel injector is very important.

Port or multipoint injection should be the absolute best for cylinder-to-cylinder distribution, but there may be problems. In this case we depend on the injector manufacturer to produce injectors that are nearly identical in delivery capability—or a matched set. The balance can also be upset if one or more injectors become partially clogged.

Manifold design also influences vaporization because the passage size affects mixture velocity and heating. If the mixture travels slowly, some liquid fuel particles will deposit onto the manifold walls before they have a chance to vaporize. The hot-spot size and location and the surface area inside the manifold influence vaporization.

When vaporization is poor, an excess of liquid fuel gets into the cylinders. Because it does not completely burn due to the lack of time available for evaporation and burning, it is expelled as unburned hydrocarbons. Excess fuel can wash oil off of the cylinder walls to cause rapid wear. A portion of this liquid fuel drains past the rings into the crackcase where it dilutes the oil.

Exhaust-Heated Hot Spots– These spots are typically small areas just under the area fed by the carburetor. Manifold ends are not usually heated. The hot spot is kept as small as possible, consistent with the needs for flexible operating and smooth running. Because the spot is small, the manifold automatically cools as rpm is increased. The large amount of fuel being vaporized at high speed extracts heat from the manifold, often making it so cold that water condenses on its exterior surfaces.

Although most passenger-car manifolds heat the mixture with an exhaust-heated spot, some are water-heated by engine coolant. Cars equipped with emission controls often heat the incoming air by passing it over the exhaust manifold on its way to the air cleaner inlet.

Intake Manifolds–Excepting racing intake systems, intake manifolds are compromises. Their shape, cross-sectional areas and heating arrangements accomplish the necessary compromises between good mixture distribution and VE over the range of speeds at which the engine will be used.

If only maximum or near-maximum rpm is being used, high mixture velocity through the manifold helps to ensure good distribution. This helps to vaporize the fuel, or at least hold the smaller particles of fuel in suspension in the mixture. At slower speeds, using manifold heat is essential to ensure vaporizing the fuel. If heat is not used, the engine will run roughly at slower speeds and distribution problems will be worsened.

Fuel Composition–The more volatile the fuel, the better it will vaporize. All fuels are blends of hydrocarbon compounds and additives. Blending is typically accomplished to match the fuel to ambient temperature and altitude conditions.

Thus, fuel supplied in the summertime has a higher boiling point than fuel available in winter. Fuel volatility is rated by a number known as *Reid Vapor Pressure*. The higher the number, the greater the volatility or tendency to go from the liquid to a gaseous state. In Detroit, for example, the Reid number varies from 8.5 in the summer to 15 in winter.

Fuel blending at the refinery is a compromise made on the basis of estimates of what the temperature will be when the gasoline is used at the pump. A sudden temperature change, such as a warm day in winter, usually causes a rash of vapor-lock and hot-starting problems.

Carburetor Placement–Location of the carburetor on the manifold in relation to the internal passages can drastically affect distribution. If the geometric layout of the manifold places the carburetor closer to one or more cylinders, this can create problems. Carburetor location is especially important with multiple carburetors.

The engine designer checks power, economy and mixture distribution with the carburetor, air cleaner and manifold installed. Sometimes the physical positioning of the air cleaner or a connecting elbow atop the carburetor is so critical that turning it to a different position can create distribution problems.

Throttle-Plate Angle–A directional effect is given to the fuel when the throttle is partially open. This effect becomes especially apparent when there is little or no riser between the carburetor and the manifold passages. A riser is the vertical passage conducting mixture from the carburetor into the intake manifold.

On a V8 engine, throttle-plate angle affects distribution in the upper level of a cross-H, dual-plane or two-level manifold more than it does the lower level. The longer riser into the lower level has a straightening effect on the mixture.

This leads to one of the advantages

Holley high-rise manifold for Chervolet small block. Two distinct levels below carb-mounting pad identify this as Cross-H or dual-plane design. Ribs in manifold floor help distribution; they direct mixture and aid in controlling liquid fuel when vaporization is not perfect.

of high-riser designs that allow better straightening of the mixture flow with less directional effects at all throttle openings. High-riser designs typically offer better cylinder-to-cylinder mixture distribution. Riser height is usually limited by hood clearance. Holley tests have shown definite improvements in part-throttle distribution with risers 1-1/2–2-in. high before opening into the manifold branches.

Mixture Speed & Turbulence–
Mixture velocities and turbulence within the manifold affect vaporization and consequently distribution. Turbulence in the combustion chamber helps to prevent stratification of the fuel and promotes rapid flame travel.

Time–Volatility of the fuel and heat available to assist in vaporization are especially important when you consider the tiny amount of time available to vaporize the atomized fuel supplied by the carburetor. Unlike

water in a tea kettle, which can be left on the stove until it boils, fuel must be changed to the vapor state in about 0.003 second in a 12-in.-long manifold passage at a mixture velocity of 300 feet per second.

DISTRIBUTION SUMMARY

To summarize the factors that can aid distribution:

• Vaporize as much fuel as possible in the manifold so a minimum of liquid fuel gets into the cylinders.

• Use fuels with the correct volatility for ambient-temperature and altitude conditions.

• Use the smallest passages consistent with the desired VE to ensure high velocities in the manifold and cylinder head.

• Carefully select venturi size (carburetor flow capacity) to have good

atomization in the carburetor.

• Avoid manifold construction that causes fuel to separate out of the mixture due to sharp turns and severe changes in cross-sectional area.

• Have sufficient turbulence in the manifold to ensure fuel and air are kept well mixed as they travel to the cylinder.

• Supply manifold heating and heating of the incoming air to ensure that fuel is well vaporized.

Not all of these factors can be controlled by an individual. But, they point out a few important considerations, such as avoiding carburetors that are too large for the engine. Use fuel that is correct for the season; don't try to race with fuel you bought in a different season or another locality. And, for engines being operated on the street, use a heated manifold with the smallest passages consistent with desired performance.

How The Carburetor 2
Works

Understanding how your carburetor works is the key to extracting top performance from your engine/carburetor combination. It's a quick way to get more performance with less work. You'll understand its internals and then do precise tuning without wasting time. Tuning by trial and error can wreck a carburetor, leave an engine in a worse state of tune and cost you money.

You'll be able to choose the right carburetor for your driving application without having to rely on guesswork or rumors about the current "hot setup."

There's really nothing tricky about how a carburetor works. No black magic makes one work differently from another. All carburetors operate essentially the same way.

Just as you can learn about an engine by examining the operation of one cylinder, we'll start by examining a one-barrel carburetor.

After you've become familiar with the one-barrel, then we'll proceed to two- and four-barrel units which are merely several one-barrels built into a single unit. It's essential to learn the basics of a one-barrel before trying to comprehend more complex carburetors. Reading this chapter first will help you establish a solid background.

A carburetor:

• Controls engine input and therefore controls power output.

• Mixes fuel and air in the correct proportions for engine operation.

• Atomizes and vaporizes the fuel/air mixture to put it in a state for combustion.

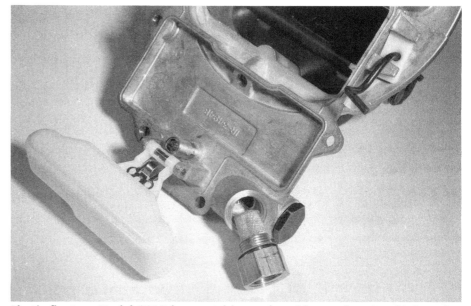

Plastic float on Model 2010 has metal insert that rides against tip of inlet valve (arrow). Float has been laid back to expose tip of inlet valve. Brass wire-mesh filter in inlet fitting catches dirt which could cause inlet-valve malfunction. Black plug closes a second inlet that can be used as an alternate fuel inlet or for a fuel-return line to the tank to purge vapors from fuel and to help cool fuel to avoid vapor lock.

Let's look at a carburetor's basic systems one at a time and see what parts do each job. We'll show the relationships involved and how they all work together. After reading this material you will be well on the road to a complete understanding of how any carburetor works.

INLET SYSTEM

The inlet system consists of three major items:

• Fuel bowl
• Float
• Inlet valve (needle and seat)

Fuel for the basic metering systems is stored in the *fuel bowl*. The fuel-inlet system maintains the specified fuel level in the bowl. Basic fuel-metering systems are calibrated to deliver the correct mixture only when the fuel is at this level. Correct fuel level also greatly affects *fuel handling*, which is the carburetor's ability to withstand maneuvering: Quick accelerations, turns and stops.

Amount of fuel entering the bowl through the fuel-inlet valve is determined by the space between the movable needle and its fixed seat (*flow area*), and by pump pressure. Needle movement in relation to the seat is controlled by the float, which rises and falls with fuel level. As fuel level drops, the float drops to open the needle valve, allowing fuel to enter the bowl. When the engine is running with constant load, the float moves the needle to a position where it restricts the flow of fuel. Only enough fuel is admitted to replace that being used. Any slight fuel-level change causes a corresponding float movement. Opening or closing the fuel-inlet valve restores or maintains the correct fuel level.

DESIGN FEATURES

Fuel Bowl—The fuel bowl or float chamber is a reservoir supplying all fuel within the carburetor.

Two- and four-barrel staged carburetors have primary throttles that open first; secondary throttles open later when higher engine power is required. Primary and secondary may have separate float bowls.

Fuel height in the bowl is controlled by the float and inlet valve. An air passage or vent connects the float bowl to the carburetor's air-inlet passage.

Venting the fuel bowl to outside air means the fuel is no longer pressurized by the fuel pump. It is pressurized only up to the float valve. Once it is in the fuel bowl, it is at *vented pressure*, the same as outside air.

The bowl is a vapor separator. Vapors entrapped in the fuel as it was pumped from the tank escape through this vent so pressure does not build in the bowl.

Connecting the vent to the inlet air horn vents the fuel bowl to clean air obtained through the air cleaner. Because the vent "sees" the same pressure as the carburetor inlet, a dirty air cleaner does not change the F/A ratio. Dirty air cleaners restrict airflow, create lower absolute pressure to the carburetor, and cause power loss. If the fuel bowl is not vented to the inlet air horn, a dirty air cleaner will richen the mixture.

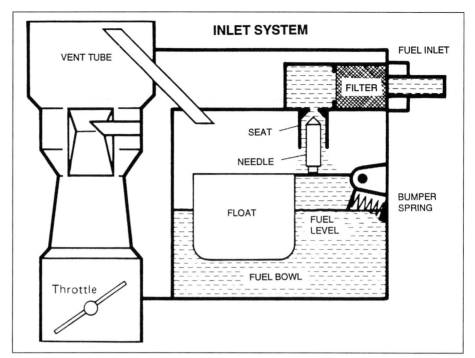

Inlet system schematic. Float has not reached desired level, so inlet valve is admitting fuel to bowl. Fuel flow shuts off when float rises to close inlet valve needle against seat.

Three Holley removable fuel bowls. Left is center-hung, center-pivot or center-inlet "race" type. It can be tapped for fuel entry on either or both sides. Center is side-hung bowl. Both bowls have externally adjustable inlet valves to set float and fuel level. Adjusting nut positions inlet valve; lock screw holds the adjustment. "Nose" bowl at right is available for restorations of older Shelby Mustangs and real Cobras. These are supplied as the LeMans Fuel Bowl Conversion Kit 34-14.

The preferred vent location is near the center of the fuel bowl. It must be high enough so fuel will not slosh into the air horn during hard stops or maneuvering. 1970 and later cars have an additional vent that connects the float bowl to a charcoal canister when the engine is turned off.

This emission-control system collects escaping fuel vapors boiled off when heat from the engine block warms gasoline in the carburetor. The same canister may also collect fuel-tank vapors. Vapors stored in the canister are drawn into the intake manifold the next time that the engine is

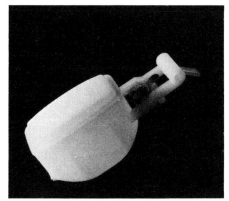

In 1987 Holley introduced hollow plastic floats as direct replacements for many brass and nitrophyl floats.

Side-hung float in secondary fuel bowl shows hollow plastic float (1), adjustable inlet-valve assembly (2), adjusting nut (3), lock screw (4), plastic baffle (5) to direct inlet fuel and contain froth, float-bumper spring (6), sight plug (7) for fuel-level checking, hole (8) for transfer tube that directs fuel from the primary to the secondary bowl. Fuel-inlet fitting is (9). When correctly set up, half-round float is adequate for all but the most-severe cornering loads such as generated on a flat asphalt track, road racing or autocrossing.

running. Pre-emission-control models often had a mechanical vent valve on the fuel bowl. At curb idle or when the engine was stopped, this external vent released fuel vapors into the engine compartment.

The fuel bowl may be an integral part of the main body casting or it can be a separate casting attached to the carburetor body with screws. Holley uses both designs and each has a replaceable screw-in inlet valve.

Bowl capacity is important. It must contain sufficient fuel to allow good response when accelerating from stop after a hot engine has been idling or stopped (called a *hot soak*). Under these conditions the fuel pump may deliver intermittent spurts of liquid fuel and pockets of fuel vapor.

Thus the bowl must be large enough to supply fuel to the metering systems until the pump is purged of vapor and able to deliver liquid fuel. Holley carburetors typically have large-capacity bowls.

Bowl capacity may be reduced with bowl inserts or *stuffers* to allow the vehicle to pass evaporative-emission requirements.

Float—A float on a hinged lever operates the inlet valve so fuel enters the bowl when the fuel level is below the desired reference height. It shuts off fuel when the desired height is reached.

A float may have one or two buoyant elements, called *lungs*. Their shape is often similar to the human lung.

Some floats are made from brass stampings soldered into an air-tight assembly, or they are made from a closed-cell plastic material not affected by gasoline, alcohol or any other commonly used fuel or fuel additive. In 1987 Holley began using a molded hollow-plastic float in many applications.

The float is designed with buoyancy to shut the inlet valve positively when the desired fuel level has been reached. The buoyant portion is usually mounted on a lever to multiply the buoyancy effects of the float. The lever also provides a surface for operating the inlet valve.

Float vibration is caused by engine or vehicle vibration or bouncing—or both. It causes wide fuel-level variations because a vibrating float allows the inlet valve to admit fuel when it is not needed. A float spring is sometimes added under the float or on a tang to minimize float vibration. These are called *float bumper springs*. Some carburetors use a tiny spring inside the inlet valve instead of, or in addition to, a spring under the float. Such springs are especially helpful in dirt-track, off-road and marine applications.

Float shape and mounting (*pivot orientation*) may be dictated by bowl size and carburetor use. The float has to provide enough buoyancy to close the inlet valve. Sometimes this requires making an odd-shape float to get enough volume for the required buoyancy.

Float-lever length determines the mechanical advantage that float buoyancy (and float spring, if used) can apply to close the inlet valve at a given fuel level.

Every carburetor has a published *float setting*. The setting is the float location which closes the inlet valve when the required fuel level in the bowl is reached.

In many Holley performance and replacement carburetors, a threaded inlet-valve assembly can be adjusted to set fuel level without taking the carburetor apart. A sight plug in the fuel bowl is removed to observe fuel level.

Inital float settings are established by measuring float position, usually in relation to a gasket surface or the top of the bowl while the carburetor is disassembled.

EFFECTS OF INLET-SYSTEM CHANGES

Change	Effect
High float level	Raises fuel level in bowl. Speeds up main system start-up because less *depression*—reduction of pressure—is required at venturi to extract fuel from bowl. Start of flow from the bowl via the nozzle is sometimes called *pullover*. Increases fuel consumption. May cause fuel to spill through discharge nozzles and vent into carburetor air inlet on abrupt stops or turns to cause over-rich air/fuel mixture. Engine then runs erratically or stalls. This spillage affects low rpm emissions. Increases chances for the carburetor to *percolate* and *boil over*. This is a condition in which fuel is pushed by rising vapor bubbled out of the discharge from the main well when it is hot. Also, when the vehicle is parked on a hill or side slope, high float level may cause fuel spillage through the vent or main system.
Low float level	Lowers fuel level in bowl. Delays main system start-up because more vacuum is required at venturi to start fuel flow. This delay may cause flat spots or *holes* in power output. May expose main jets in hard maneuvering, causing *turn cutout*—misfire from lean air/fuel mixture. Can lessen maximum fuel flow at wide-open throttle—WOT.
Float assist spring too strong	Causes low fuel level. Inlet valve closes prematurely because of added force of spring.
Assist spring too weak	Causes high fuel level. Inlet valve closes after correct fuel level is reached because more float movement is required to compensate for weak spring.
High fuel pressure	Raises fuel level approximateiy 0.020-in. for each psi—pounds per square inch—fuel pressure increase. This factor varies with bouyancy, leverage, and needle-orifice size. Don't exceed 9 psi.
Low fuel pressure	Lowers fuel level
Larger inlet-valve seat	Raises fuel level.
Smaller inlet-valve	Lowers fuel level.

NOTE: Changing almost any part in fuel-inlet system requires resetting float to maintain correct fuel level.

A fuel level for a particular carburetor is established by the designer and test engineer so the carburetor will operate without problems. Fast starts and stops and maneuvers ordinarily encountered in the particular vehicle must be considered. The level is set so there will be no fuel spillage when a passenger car is parked or operated facing up, down or sideways on a hill with 32% grade (18°).

A military-vehicle specification requires correct operation when parked or operated at 60% (31°) up/down and/or a 40% (22°) side slope.

For best operation in high-speed cornering, bowls are equipped with *center-pivoted floats*. The pivot axis is parallel with the car axles. Effects of acceleration and braking are best resisted by *side-hung floats* with pivots perpendicular to the axles.

Fuel tends to slosh-over through the vent into the throttle bores as it moves toward the carburetor throttle bores on hard stops. To offset this, fuel levels in rear bowls are typically set 1/16- to 1/8-in. lower than the forward or primary bowls. This lower fuel level makes the fuel rise higher before it can spill over through the main discharge nozzle or vents. Thus the carburetor is more tolerant of hard stops.

Inlet Valve—The needle's rounded end rests against the float lever arm, or is connected to it with a hook. The tapered end of the needle closes against an inlet-valve seat as the float rises.

Some inlet needles are hollow. They may contain a tiny damper spring and pin to help cushion the needle valve against road shocks and vehicle vibrations transmitted through the float.

Inlet-valve needles are usually steel with a tapered seating surface. Viton-tipped needles are extremely resistant to dirt and conform to the seat for good sealing with low closing forces. Plain-steel needles should be used with fuels such as alcohol and nitromethane.

Inlet valves are supplied with various seat openings. Each inlet-valve assembly (if the needle is an integral part) and each inlet-valve seat is stamped with a number. This number indicates seat opening in thousandths of an inch. A valve assembly marked 110S has a 0.110-in.-diameter seat opening and a Viton-tipped needle; a 110 is the same seat opening with a steel needle.

Seat diameter is an indication of how much fuel flow occurs at a given fuel pressure. A smaller opening flows less fuel—a larger opening more. The chart on page 24 shows fuel flow with a 0.110-in. inlet seat at various fuel pressures with the needle fully open (float dropped all the way down).

Seat size is selected to allow reasonably quick bowl filling to handle quick accelerations after standing parked with a hot engine. Minimum restriction is also needed for high fuel demands such as wide-open throttle (WOT) at high rpm. Larger seats provide better vapor purging from the fuel lines. A small needle seat offers the best hot-fuel control because vapor pressure in the fuel line acts against less area to force the needle

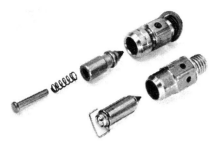

Two spring-loaded inlet-valve assemblies. Top unit has spring to cushion needle when float bounces needle upward against seat. This type of needle is used in Holley's side-hung off-road fuel bowls, Part 34-3. Hooked needle at bottom has captured spring and ball end in a hollow needle.

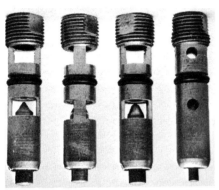

Round-hole inlet-valve is used for low-flow applications. "Picture-window" assemblies are common where higher flow is required. O-ring seals inlet side from fuel bowl. Threaded portion allows fuel-level adjustment.

off of its seat. A too-large needle seat is a definite hindrance for off-road and other types of racing. Because a small needle seat aids fuel control, always use the smallest-possible inlet valve that will work with the engine. Any needle-seat change requires resetting the float to the float-height specification.

Fuel Filter—A fuel filter or screen may be included in the carburetor body, fuel-bowl or cover as part of the fuel-inlet system. The filtering device is placed between the fuel pump and the inlet valve to trap dirt which could

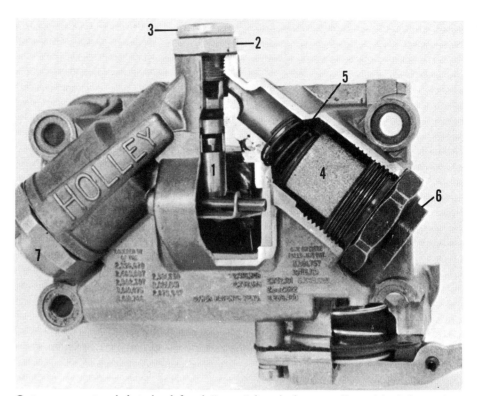

Cutaway center-inlet dual-feed "race" bowl shows adjustable inlet-valve assembly (1), adjusting nut (2) and lock screw (3). Sintered-bronze filter (4) in inlet has spring (5) to allow fuel bypass if filter plugs up. Fuel-inlet nut (6) can be swapped with plug (7) to allow plumbing opposite side. Inside view is shown below.

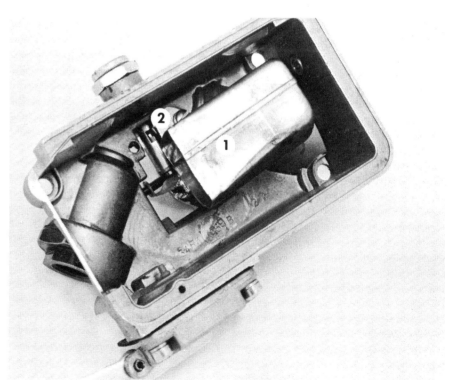

Cutaway side view of dual-feed "race" bowl shows half of float (1) and segment of mounting bracket (2).

FUEL FLOW vs. FUEL PRESSURE FOR VARIOUS SIZE NEEDLES & SEATS

Dia. of Needle Seat	Type	Fuel Flow @ 2 P.S.I lbs./hr.	Fuel Flow @ 4 P.S.I. lbs./hr.	Fuel Flow @ 6 P.S.I. lbs./hr.
0.082″	Holes	106	153	204
0.097″	Holes	121	174	225
0.101″	Holes	138	194	254
0.110″	Holes	153	230	275
0.110″	Windows	160	232	295
0.120	Windows	167	236	305

NOTE: Checked with needle & seat assembly installed with float held at bottom of bowl.

Holley Tests
October, 1971

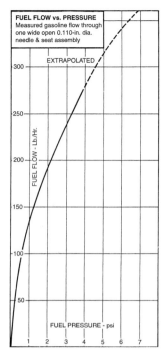

Measured gasoline flow through one wide-open 0.11-in.-diameter window-type inlet valve assembly. Float held at bottom of bowl.

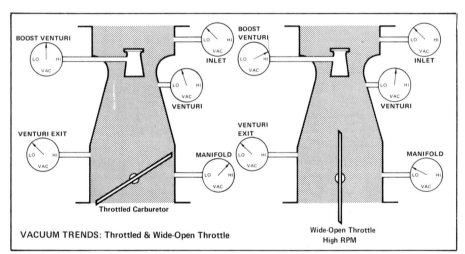

Typical carburetor vacuums inside carburetor air section. Slight vacuum at inlet represents pressure drop across air cleaner. Gauge at large venturi throat shows higher vacuum because air is still at relatively high velocity. Vacuum returns almost to inlet value just before throttle plate. Throttled carburetor shows very high manifold vacuum because a large pressure drop occurs across the partially opened throttle. Wide-open carburetor's low manifold vacuum indicates a heavy-load/dense-charge condition. In both cases, highest carburetor vacuum is at the boost venturi throat which supplies signal to main system.

cause inlet-valve seating problems. Fuel filters are discussed on page 210.

MAIN SYSTEM

The main metering system supplies fuel/air mixture to the engine for cruising speeds and above. Fuel is fed by the *idle* and *accelerating-pump* systems until engine speed or airflow increases to a point where the main metering system begins to operate. When the engine must produce full power under high-load conditions, added fuel comes from the *power system*. These other carburetor systems are explained elsewhere in this chapter.

Throttle—Many people believe the throttle controls the *volume* of fuel/air mixture being pumped into the engine. Not so! Piston displacement never changes, so the *volume* of air pulled into the engine is constant for any given speed. The throttle controls the *density* or *mass flow* of the air pumped into the engine by piston action: *Least* charge density is available at idle, *highest* density is at WOT. A dense charge has more air mass, therefore higher compression and burning pressures can be developed for higher power output. So, the throttle controls engine speed and power output by varying the *charge density* supplied to the engine.

Venturi—Let's proceed to the heart of the carburetor, the venturi. This is the one simple part that really makes a carburetor function. It is important to understand how the venturi operates before looking further into how the main metering system works.

The venturi and its principle of operation are named after G. B. Venturi, an Italian physicist (1746-1822). He discovered when fluid flows through a constricted tube, flow is fastest and pressure lowest at

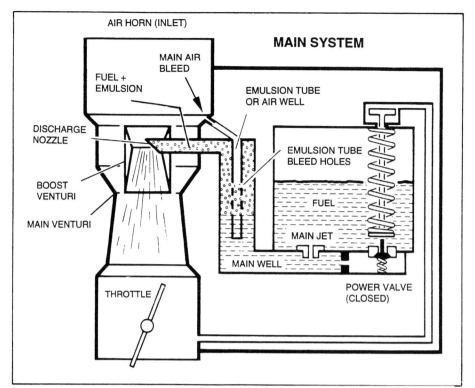

Main venturi's reduced diameter speeds up airflow. Reduced pressure in venturi area allows air stream to pick up fuel flowing from bowl through passages. Main jet in fuel path limits fuel flow.

the point of maximum constriction in the tube.

In the internal-combustion engine, a partial vacuum is created in the cylinders by the pistons' downward strokes. Because atmospheric pressure (14.7 psi at sea level) is higher than the reduced pressure in the cylinders, air rushes through the carburetor and into the cylinders to fill the vacuum. On its way to the cylinders, air passes through the venturi.

The venturi is a smooth-surface restriction in the path of the incoming air. It constricts or "necks-down" the inrushing air column, then allows it to widen back to the throttle-bore diameter. Air is rushing in with a certain pressure. To get through the necked-down area (venturi), it must speed up, which reduces pressure inside the venturi. A gentle diverging section is used to recover as much of the pressure as possible.

The venturi is the controlling factor in the carburetor because fuel dis-charges into the venturi at the point of lowest pressure (greatest vacuum). This minimum-pressure point applies a *signal* to the main metering system.

The pressure drop or vacuum signal is "seen" at the discharge nozzle in the venturi. Because the fuel bowl is maintained at near-atmospheric pressure by the vent system, fuel flows through the main jet and into the low-pressure or vacuum area in the venturi.

Pressure drop (vacuum) at the venturi varies with engine speed and throttle position, increasing with engine rpm. Wide-open throttle and peak rpm give the highest fuel flow and the highest pressure difference between the fuel bowl and a discharge *nozzle* in the venturi. Pressure difference (ΔP) as engine speed changes is approximately proportional to the difference in velocity squared. (ΔV^2), $\Delta P \sim \Delta V^2$.

Pressure drop in the venturi also depends on venturi size. A small ven-turi provides a higher pressure drop at any given rpm and throttle opening than a large venturi. The Design Features portion of this section explains more about this important consideration and how it relates to performance.

No fuel issues from the discharge nozzle until flow through the venturi and hence, pressure drop, is sufficient to offset the level or "head" difference between the discharge nozzle and the lower level of fuel in the bowl.

Main Jet—Once main-system flow starts, fuel is metered (measured) through a main jet in the fuel bowl. From the main jet, fuel passes into a main well. As fuel passes up through this main well, air from a main air or "high-speed" bleed is added. This pre-atomizes or emulsi-fies the fuel into a light, frothy fuel/air mixture that issues from the discharge nozzle into the air stream flowing through the venturi. The discharge nozzle is often located in a small boost venturi centered in the main venturi.

Liquid fuel is converted into a fuel/air emulsion for two reasons. First, it vaporizes much easier when it is discharged into the air flowing through the venturi. Second, the emulsion has a lighter viscosity than liquid fuel and responds faster to any change in the venturi vacuum (signal from the venturi applied through the discharge nozzle). It will start to flow sooner and quicker than liquid fuel.

Main Air Bleed—The strong signal from the discharge nozzle is bled off or reduced by the main air bleed so there is less effective pressure difference to cause fuel flow. The mixture becomes *leaner* as bleed size is increased. Decreasing bleed size increases pressure drop across the main jet. More fuel is pulled through the main system, giving a *richer* mix-ture. Main air-bleed changes affect the entire range of main-metering-system operation. Holley establishes main air-bleed size for each carburetor to work

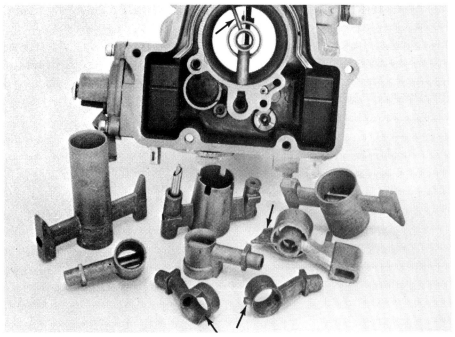

Boost venturis may have wings or tabs (arrows) to make a carburetor work on a specific engine/manifold combination. Picture how these work by touching the edge of water streaming from a faucet. Mixture is deflected in the same way.

correctly over that carburetor's airflow range. Changes in main-air-bleed sizes are rarely necessary or advisable. Calibration changes are easy to make by changing main metering jets.

The main air bleed also acts as an *anti-siphon* or *siphon-breaker* so fuel does not continue to dribble into the venturi after airflow is reduced or stopped.

DESIGN FEATURES

Throttle—The throttle shaft is offset slightly (about 0.020 in. on primaries and about 0.060 in. on secondaries) in the throttle bore. This gives one side of the throttle a larger area to cause self-closing.

There are two reasons for this design. First, idle-return consistency is greatly aided by the sizable closing force generated when manifold vacuum is high—as at idle. Second, it is a safety measure to guard against over-speeding an engine started without installing the linkage or throttle-return spring. Airflow past the throttles will tend to close them.

Throttles seldom close tightly against the throttle bore; they are factory-set against a stop. The factory setting provides a closed-throttle airflow specified for that particular carburetor.

Venturi—The most efficient (or *ideal*) venturi that creates maximum pressure drop with minimum flow loss requires a 20° entry angle and a diverging section with a 7° to 11° included angle on the "tail." The "ideal" venturi will not always fit into a carburetor short enough to fit under the hood of the average automobile.

Holley uses a radiused entry to the venturi because it is less affected by production variations. Designers try to keep as close as possible to the ideal venturi entry and exit angles within the limitations imposed by engine position and hood height.

Although the theoretical low-pressure point and point of highest velocity would be expected at the venturi's minimum diameter, fluid friction causes the point to occur about 0.030 in. below the smallest diameter. This low-pressure, high-velocity point is called the *vena contracta*. The center of the discharge nozzle or the *tail* of the boost venturi is placed at this point.

A venturi allows a much greater metering signal than a straight tube and has a minimum airflow loss. This is because the venturi's trailing edge conforms to the normal airstream. It recovers most of the pressure drop so a greater mass of air is available for the engine. It is a very efficient air-metering device because of the high signal levels it provides with minimum pressure loss.

The venturi supplies fuel in the correct proportion to the mass of air rushing into the engine. Its size affects the pressure drop available to operate the main-metering system. The *smaller* the venturi, the *greater* the pressure drop, the *sooner* the main system will be brought into operation, and the *better* the mixing of fuel with air. The rpm at which main system spill-over or *pull-over* starts is affected by the size of the engine that is pulling air through the carburetor.

With any given engine size, a larger carburetor requires a higher rpm to bring the main system into operation; a smaller carburetor, a lower rpm. But venturi size also controls the amount of air available at wide-open throttle (WOT). If the venturi is too small, top-end HP will be reduced, even though the carburetor provides very good fuel/air mixing at cruising speeds. For this reason automobile manufacturers typically compromise.

They use carburetors with smaller-than-optimum (for maximum power) airflow capacities. And for very good reasons. Good fuel atomization and vaporization promote good distribution and improve running and economy at around-town and cruising speeds.

Manufacturers offering more performance typically supply a carburetor with a secondary system. This retains the advantages of a small primary venturi with the capability of higher airflow for top-end power.

Air from main air bleed can be introduced into main well in several ways. Sometimes it is brought in through the center of an emulsion tube. Separate emulsion tubes are from a Holley 2110 Bug Spray (right) and a 5200 (left). In carburetors with removable metering blocks (plates, bodies), bleed air is often introduced through tiny holes in an air well paralleling the main well. This well is on the surface assembled toward carburetor body. Arrows indicate bleed holes which are necked down to about 0.026-in. diameter.

Boost Venturi—These act the same as the larger venturi, but supply a stronger signal to the discharge nozzle because boost-venturi velocity is higher than that in the main venturi. Boost venturis are *signal amplifiers*.

By increasing the available signal for main-system operation, the boost venturi allows the carburetor to work well at lower speeds, and therefore lower airflows. This is especially helpful in a performance-type carburetor because the boost venturi does not seriously affect the carburetor's airflow capacity.

The boost-venturi tail discharges at the low-pressure point (*vena contracta*) in the main venturi. Consequently, boost-venturi airflow is accelerated to a higher velocity because it "sees" a greater pressure differential than the main venturi.

Air and fuel emerging from the boost venturi are traveling faster than the surrounding air. A shearing effect between the two airstreams improves fuel atomization.

Boost venturis also aid fuel distribution because the ring of air flowing between the two venturis directs the charge to the center of the airstream. This helps to keep some of the wet fuel/air mixture off of the carburetor wall below the venturi so more of it reaches the manifold hot spot for improved vaporization.

Tabs, bars, wings and other devices may be used to direct the mixture flow to improve cylinder-to-cylinder distribution with a particular manifold. This development work is always accomplished on the dynamometer during design and development of the carburetor/manifold combination.

Boost venturis allow using a much shorter main venturi so the carburetor can be made short enough to fit under the hood. Carburetor designers could achieve essentially the same results with a long (ideal) venturi as they can with one or more boost venturis stacked in the main venturi. But it is tough to build a carburetor which allows the 20° entry to the venturi and a 7° to 11° "tail," and at the same time get venturi size down to the required diameter for adequate signal. Some Holley carburetors, especially very short ones, use two booster venturis to get adequate signal for main-system operation.

Main Jets—These metering orifices control fuel flow into the metering system. They are rated in flow capacity and are removable so the mixture can be changed.

Main jets are not selected solely by trial and error. For a given venturi size, a small range of main jets will cover all conditions. Nevertheless, final selection of the correct jet for the

Look-alike main jets. Only the markings indicate a difference. 50 is standard jet. Two-digit (50) series jets have 3% possible variation between any two jets with the same marking. Sizes are available from 40 through 100. Order 122-XX (jet number). Close-limit series 501, 502, 503 have only 1.5% flow difference (maximum) between any jet with the same marking. Close-limit main-jet sizes from 352 to 742 are available. Order 122-XXX. Jets with numbers ending in 1 or 3 are used by the factory during calibration.

application has to be done by testing. Design and operational variations (climate, altitude and temperature) affect jet-size requirements.

There's a basic misconception about jets: that size alone determines flow characteristics. Actually, the shape of the jet entry and exit, as well as the finish, affect flow. Holley checks each jet on a flow tester and grades it according to flow. This flow rate is compared to a master chart. A number is stamped on the jet to indicate its flow. The tolerance range for each size explains why a 66 jet may not seem to give a richer mixture than a 65. If the 65 is on the high side of its allowable tolerance and the 66 is on its low side, then the two jets may flow very close to the same amount of fuel. The 65 tolerance range is from 351.5—362.0 cubic centimeters per minute at a specified head with a given test fuel. The 66 range is from 368.5—379.5 cc per minute.

In each case, there is a 3% flow range within a jet size, and about 4.5% difference in flow between average jets in the two sizes. In 1975 Holley developed a new *close-limit* series of main jets in the standard size range of 30 to 74. This new series was developed so

27

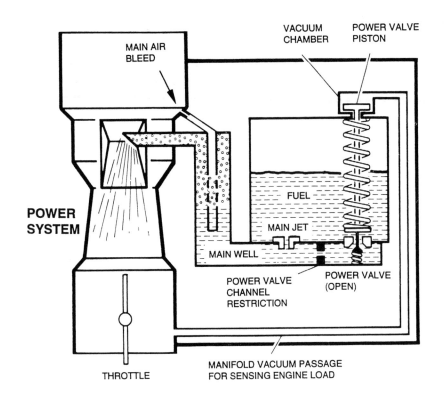

POWER SYSTEM

MAIN AIR BLEED

VACUUM CHAMBER

POWER VALVE PISTON

FUEL

MAIN JET

MAIN WELL

POWER VALVE CHANNEL RESTRICTION

POWER VALVE (OPEN)

THROTTLE

MANIFOLD VACUUM PASSAGE FOR SENSING ENGINE LOAD

POWER SYSTEM

DISCHARGE NOZZLE

BOOSTER VENTURI

MAIN AIR BLEED

POWER VALVE CHANNEL RESTRICTION

POWER VALVE

MAIN JET

MAIN WELL

MANIFOLD VACUUM

Typical vacuum and fuel-passage routing used in Holley carburetors with removable fuel bowls. These are diaphragm-type power valves. Using a power valve on the secondary side allows using a smaller main jet to improve fuel handling during deceleration and braking.

Typical Holley "window" power valve at right is used in Models 2010, 2300, 2305, 3150, 4010/11, 4150/60/80, 4165/75 and 4500. Only the window valve is currently available and it is offered in two sizes: 125-XX for carburetors with two power-valve-channel restrictions up to 0.095-in.; 125-165 and 125-1005 for larger PVCRs. Drilled-hole valve (left) was obsoleted in the mid-'80s, but continued to be used in some factory-assembled carburetors after that time.

flow could be closely tailored to fit emission requirements.

These jets use the first two numbers, such as 65 or 66, with a third number to indicate whether the jet flows lean (651) in the middle (652), or rich (653). There is approximately 1.5% difference in flow between each of the three jets, or a flow range of 4.5%. There can only be a 1.5% difference in flow between two jets with the same flow marking. Jets with the same markings in the old two-digit numbering system could have up to 3% flow difference.

These jets are the same brass or aluminum color as used previously.

Holley offers close-limit jets in mean sizes only, that is, 352, 262, etc. You may have a carburetor with a 351 or a 353 in it, but these jets are only available in the assembly plant where the carburetor is made and flow-checked.

Close-limit jets offer a good way to accomplish fine tuning, provided you can use jets ranging from 35 to 74 (352 through 742). Order these jets as 122-XX2. Consult the current

Holley Performance Parts Catalog for available sizes.

Don't drill jets to change their size. This always destroys the entry and exit features. It may introduce a swirl pattern, even if the drill is held in a pin vise and turned by hand. You cannot be sure of the flow characteristics of a jet that has been modified by drilling.

Main Air Bleed—All Holley high-performance carburetors are equipped with built-in fixed-dimension air bleeds. A flat surface in the inlet horn is used for mounting these bleeds. They see total pressure and are not affected by airflow variations.

POWER SYSTEM

When the engine is called upon to produce power in excess of normal cruising requirements, the carburetor has to provide a richer mixture. This is discussed in the fuel-requirements section. Added fuel for power operation is supplied by the power system that is controlled by manifold vacuum.

Manifold vacuum accurately indicates engine load. Vacuum is usually strongest at idle. As load increases, the throttle valve must be opened wider to maintain a given speed. This offers less restriction to air entering the intake manifold and reduces manifold vacuum.

A vacuum passage in the carburetor applies manifold vacuum to a power-valve piston or a diaphragm in a screw-in power valve. At idle or normal cruising-load conditions, manifold vacuum acting against a spring holds the valve closed. As high power demands load the engine, manifold vacuum drops.

Below a preset point, usually about 6 inches of mercury (in.Hg), the power-valve spring overcomes manifold vacuum and opens the power valve. Fuel flows through the power valve and through a power-valve-channel restriction (PVCR) to join fuel already flowing through the main metering system from the main jet. The mixture is richened.

The power valve is a "switch" operated by manifold vacuum. It is designed to operate (add more fuel) at a given load. It turns on extra fuel to change from an economical cruising fuel/air mixture to a power mixture. The power-valve channel restriction (PVCR) controls the amount of enrichment.

When engine power demands are reduced, increasing manifold vacuum acts on the diaphragm or piston to overcome power-valve-spring tension, closing the power valve and shutting off the added fuel supply.

DESIGN FEATURES

Power Valves—Holley high-performance carburetors usually have single-stage power valves. Carburetors for 'street use, especially on 1973 and later engines with exhaust-gas recirculation, have two-stage power valves. Two-stage power valves are discussed on pages 29-30.

Single-stage power valves are used in the 2010, 2300, 4010, 4011, 4150,

Numbers stamped onto flats of screw-in power valves indicate manifold vacuum at which valve opens. These show 50, 65, 75 and 85, specifying 5.0, 6.5, 7.5 and 8.5 in.Hg, respectively. Current power valve identification is stamped on a metal disc on the vacuum side of the valve. Power valves are "switches" that turn on fuel which is metered through the PVCRs to supply additional fuel for power enrichment.

4160, 4165, 4175 and 4500 series. These screw-in type valves are available with different flow areas and opening points. The flow area used must always be larger than the combined areas of the PVCRs.

Opening points are available in increments from 2.5–10.5 in.Hg. The last digits in the part number indicate the opening point in in.Hg for that single-stage power valve when a decimal point is placed ahead of the last digit. This number is also stamped on the valve. Examples:

125-105 opens at 10.5 in.Hg
125-25 opens at 2.5 in.Hg

Power-valve-opening point is another variable the carburetor designer uses to arrive at the best compromise between economy, exhaust emissions and drivability. On normal replacement carburetors, the valve may be opened quite late to allow a particular engine to meet emission requirements, as well as to maintain a broad economy range.

Racing engines, on the other hand, have substantial manifold-vacuum fluctuations at idle and low speeds. The tuner must install a power valve that won't open and close in response to variations caused by valve timing instead of throttle position or engine

load. Selecting power valves to meet these special requirements is explained on page 129.

Most Holley two- and four-barrel performance-type carburetors use *picture-window* power valves with flow capacity sufficient for two 0.095-in. PVCRs.

Some Holley power valves use 4 or 6 drilled holes in place of the picture-window type. No longer available as service parts, they may be factory-installed in some carburetors. All two-stage power valves use the drilled-hole construction.

Screw-in power valves may be stamped with additional numbers, such as A5. This manufacturing code indicates the month and year in which the valve was made. A is January and 2 is 1992. C0 indicates March, 1990. Of course, the year can be off in increments of 10 years, so it is helpful to have some idea of when the carburetor was made.

Two-stage power valves are used on replacement carburetors to allow a particular engine to pass emission tests. Staged power valves open partially at one vacuum level, then open fully when manifold vacuum falls to a lower level. The first stage in these valves opens a metering orifice smaller than the PVCR.

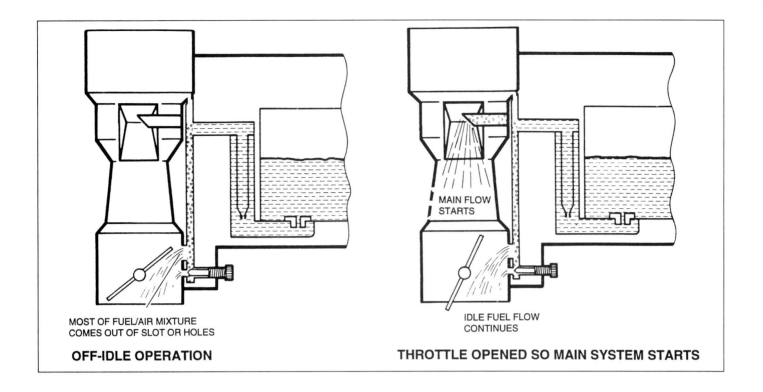

MOST OF FUEL/AIR MIXTURE
COMES OUT OF SLOT OR HOLES

OFF-IDLE OPERATION

MAIN FLOW
STARTS

IDLE FUEL FLOW
CONTINUES

THROTTLE OPENED SO MAIN SYSTEM STARTS

Holley offers two-stage power valves as replacement parts. They provide improved drivability in vehicles with relatively low power-to-weight ratios. Two-stage power valves for Models 2010, 2300, 4010, 4011, 4150, 4160 and 4180 are identified with numbers that can be deciphered with the help of a Holley catalog.

Two-stage power valves, 125-165 and 125-1005, open at 6.5 and 10.5 in.Hg respectively. These flow adequate fuel to handle two 0.128-in. PVCRs. Because of its size, the valve stem cannot be piloted and the power valve can be damaged by dirt in the fuel. Don't use them unless you have a real need.

Holley does not guarantee economy improvements, but some benefit should be realized if two-stage power valves are used as follows:

• Do not use the valves in any vehicle that will be used on the drag strip. Two-stage valves are restrictive and could cause leaning out at high rpm.

• Best economy improvement will be realized in recreational vehicles equipped with the 0-6619, 0-6909 and similar carburetors, and in some station wagons and heavy sedans using Model 4011 or 4165 spread-bore carburetors. Use the valve only on the primary side. Best results are obtained in stop-and-go driving, or driving in rolling or mountainous terrain.

• In other applications, some fuel economy may be lost. Typical two-stage power valves operate as follows: 125-206 opens the 1st stage at 12.5 in.Hg, the 2nd stage at 5.5 in.Hg. For high-altitude applications above 4,000 feet, the 125-207 opens the 1st stage at 10.5 in.Hg, the 2nd at 5 in.Hg.

IDLE SYSTEM

Idling requires richer mixtures than part-throttle operation. Unless the idle mixture is richer, slow and irregular combustion will occur. This is due to the high dilution of the intake charge by residual exhaust gases that exist at idle vacuums.

The idle system supplies fuel at idle and low speeds. The idle system has to keep the engine running, even when accessory loads are applied to the engine. These include the alternator, air conditioning and power-steering pump. The idle system also has to keep the engine running against the load imposed by an automatic transmission in one of the operating ranges (Low, Drive, Reverse).

At idle and low speeds, not enough air is drawn through the venturi to cause the main metering system to operate. Intake-manifold vacuum is high because airflow is severely restricted by the nearly closed throttle valve. This high vacuum provides the pressure differential for idle-system operation.

Perhaps the easiest way to understand how the idle system works is to look at it as a tiny main metering system. Because air enters the idle system through the idle air bleed, think of the bleed as the *venturi*. Flow through this system depends on transfer-slot and/or discharge-hole area exposed by the main throttle plate and idle-mixture-screw position. Consider the discharge hole or slot and the idle screw as forming the *idle-system throttle*.

Backing out the idle screw or opening the throttle to expose more of the slot to manifold vacuum opens the idle system "throttle" so more air flows through the system. Pressure drop across the idle air bleed increases, bringing more fuel from the idle well so the mixture stays at the desired fuel/air ratio. Fuel flows through the main jet, then into a vertical idle well, past the idle-feed restriction. It then mixes with air from the idle air bleed. See drawings below.

Sometimes the idle restriction is at the bottom of the idle well. The fuel/air mixture is lifted up and across to another vertical passage. Fuel-air mixture flows down this second vertical passage and branches in two directions: through the idle-discharge passage into the throttle bore *below* the throttle valve, and to an idle-transfer slot (or holes) just *above* the throttle valve.

When the throttle valve is opened and engine speed increases, airflow through the venturi increases so the main metering system begins discharging fuel through the discharge nozzle in the booster venturi. Flow from the idle system tapers off as the main system starts to discharge fuel. The two systems are designed to provide smooth gradual transition from idle to cruising speeds when *carburetor capacity* is *correctly matched to engine displacement*.

When the throttle is fully open, there is not enough vacuum at the idle port to draw any fuel-air mixture through the idle system. Idle system flow gradually ceases as the throttle is moved from idle to cruising. Naturally, flow through the main-metering system increases as idle flow decreases, so the transition is smooth—*if* the right carburetor is chosen for the engine.

When the throttle is closed, high vacuum below the throttle plate draws air and fuel through the idle system and out of the idle-discharge port. When the throttle is partially open, vacuum below the throttle plate is reduced and flow through the idle system is reduced. Some flow is out of the idle port and some is out of the idle-transfer port.

In normal driving, flow swings quickly back and forth between idle and main operation as the vehicle is accelerated, slowed by closing the throttle, idled at stop, and then reaccelerated.

DESIGN FEATURES

Throttle Stop—The throttle lever is seated against a stop instead of closing the throttle against the throttle bore. This helps to keep the throttle from sticking in its bore. It also makes the idle system less sensitive to mixture adjustments. Holley carburetors are factory-set to an idle-airflow specification. Do not readjust them to seat the throttle in the bore—(especially diaphragm-operated secondaries).

Idle Air Bleed Size—Increasing idle air bleed size reduces pressure drop across the bleed, reducing the amount of fuel pulled over from the idle well. Increasing idle air bleed size leans the idle mixture, even if the idle-feed restriction is left constant. Conversely, decreasing idle air bleed size increases pressure drop in the system and richens the idle mixture.

Idle Speed Setting—Before emission-control requirements became important, idle setting was typically the slowest speed at which the engine would keep running smoothly. Emission requirements make higher idle speeds necessary. A higher idle speed reduces exhaust-gas dilution which occurs at lower idle speeds so leaner idle mixtures can be used without misfiring.

Engines designed to pass emission requirements are typically set for *lean best idle* at a specified rpm and a subsequent reduction in idle speed by leaning the mixture still further—as stated on a label in the engine compartment.

Manufacturers have carefully correlated this lean best idle and subsequent idle drop-off to provide the required carbon-monoxide percentage

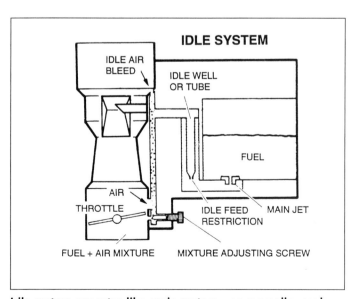

Idle system operates like main system—on a smaller scale.

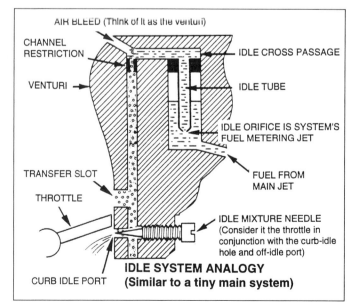

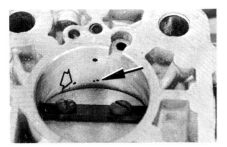

Some idle-transfer circuits (progression or off-idle) use a slot as shown above. Others use a series of holes. Curb idle is always hole nearest the flange. Outline arrows indicate spark ports.

IDLE LIMITER CAP

MAIN METERING JET

Idle-limiter caps cover screw access openings or fit on adjustment screw. Later carburetors have the adjustment screws hidden by a lead plug or by a hardened-steel plug to discourage or prevent tampering.

Cutaway Model 4500 shows intermediate idle-system discharge tube. System is fed from its own restriction in fuel bowl.

(CO%) to pass the emission requirement. From this lean-best-idle point, and at this specified rpm figure, each idle mixture-adjustment screw is turned in to provide a specified rpm drop-off. Where there are two screws, each is adjusted to provide 1/2 of the specified idle drop-off. CO levels are usually stated for use when an analyzer is available.

Older non-emission-controlled cars and racing cars are typically set for the desired idle rpm and best manifold vacuum. This is not a minimum-emission setting.

Idle Limiter—An idle-limiter cap limits idle mixture-screw adjustment to approximately 3/4 turn. Applied after the desired idle mixture has been set, this limiter prevents easy tampering with the idle-mixture adjustment. This factory setting is made when the carburetor is flow-checked. The limiter is constructed so removing it requires destroying the cap, thereby showing that the carburetor has been readjusted and may not be providing required emission performance.

Beginning in the late '70s, idle

mixture adjustment capability was virtually eliminated. Idle adjustment is trapped in a fixed setting by a limiter cap. Or, the idle-adjustment is hidden beneath a plug that is difficult or impossible to remove.

Intermediate Idle System—In five of the largest Holley high-performance carburetors (Model 4500), an intermediate idle system discharges through a tube into the venturi's trailing edge. This extra idle system provides additional transfer fuel between idle- and main-system operation. Opening the throttle past the usual idle-system transfer slot greatly reduces the manifold vacuum that was being applied to the idle system. In these large-venturi carburetors, velocity through the venturi is not sufficient to establish main-system flow at this time. A giant flat spot would occur without this intermediate idle system.

The discharge tube for the intermediate idle system is above the transfer slot at the main venturi tail. Operation of the intermediate system starts as the throttle moves past the intermediate discharge. See photo above.

The intermediate idle system has its own idle air bleed and idle feed restriction. There are no adjustments. And, the bowl-fed system continues feeding fuel after the main system starts because the discharge tube is at a lower pressure than the fuel bowl.

Auxiliary Air Bleeds—These are sometimes used in the idle system. Although they usually add air to the idle system downstream from the traditional idle air bleed, they act in parallel with the idle air bleed.

ACCELERATOR PUMP

The accelerator pump:

• Makes up for fuel that condenses onto manifold surfaces when the throttle is opened suddenly.

• Makes up for the lag in fuel delivery when the throttle is opened suddenly, which allows more air to rush in.

• Acts as a mechanical-injection system to supply fuel before the main system starts.

32

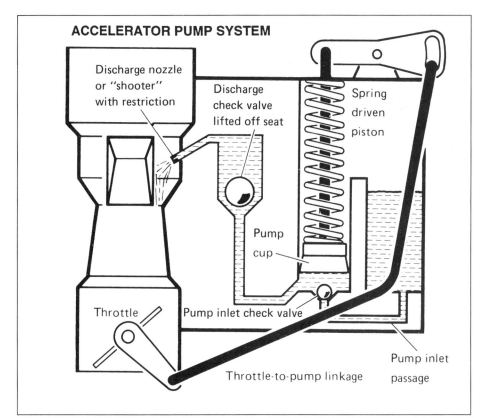

ACCELERATOR PUMP SYSTEM

Discharge nozzle or "shooter" with restriction

Discharge check valve lifted off seat

Spring driven piston

Pump cup

Throttle

Pump inlet check valve

Throttle-to-pump linkage

Pump inlet passage

Partially open throttle causes accelerator pump to move to bottom of its well. Pump inlet check ball is forced onto its seat and discharge check ball has lifted off of its seat. Fuel sprays into carburetor throat through shooter nozzle. Pump cup can also serve as the pump inlet. There is often a weight above the discharge check ball. The weight keeps the ball from lifting off its seat and flowing fuel because of strong signals at the discharge nozzle (shooter).

Two pump-inlet-valve types. Rubber umbrella-shape valve at bottom are Holley's current standard. It seats instantly for best tip-in performance. Hanging-ball type at top (now obsolete) requires an instant to seat before pump shot can be delivered. A tiny part of the shot is wasted into the bowl as pump pressure seats the ball. There's 0.011- to 0.013-in. clearance between ball and retainer when bowl is inverted.

High manifold vacuum tends to keep the mixture well vaporized. As the throttle is opened quickly, intake-manifold vacuum instantly drops, moving the pressure toward atmospheric. As pressure in the manifold rises toward atmospheric, some of the fuel drops out of the vapor and becomes liquid. It condenses into puddles and wet spots on the walls and floor of the manifold. Consequently, only a lean mixture is available for the cylinders.

The engine hesitates or stumbles unless more fuel is immediately added to replace that which fell out of the mixture. This is especially important with big-port manifolds or large-plenum manifolds because there is more surface area for fuel to condense onto.

Making up for condensed-fuel loss onto the manifold and taking care of fuel-delivery lagging behind increasing airflow are both important. A third accelerator-pump-system function is supplying fuel when the throttle is quickly opened past the point where the idle-transfer system would have supplied fuel until the main system could begin its normal operation. Here it supplies the required fuel until the main system starts flowing. During the low-flow, low-vacuum period, the accelerator pump injects fuel under pressure into the throttle bore. Duration of accelerator-pump operation must be carefully engineered to provide a "cover up" of sufficient length to allow main-system flow to be established so good vapor-

ization will be ensured and correct fuel/air ratio reestablished.

The accelerator pump operates when the pump-operating lever is actuated by throttle movement. As the throttle opens, the pump linkage operates a *pump diaphragm* or *piston*. Pressure in the pump forces the *pump-inlet ball*, valve or pump cup onto its seat so fuel will not escape from the pump into the fuel bowl. And the pressure also raises the *discharge needle* or ball off its seat so fuel discharges through a *shooter* into the venturi.

As the throttle is moved toward the closed position, the linkage returns to its original position. The pump-inlet ball, valve or pump cup moves off its seat to allow the pump to refill from the bowl. The piston or diaphragm is positively pulled back to the at-rest position. This creates a vacuum in the pump cavity to ensure quick refilling of the pump. As pump pressure is relieved, the discharge check needle or ball reseats so there is a closed check valve to refuel the pump.

The weight of the discharge-check valve keeps it closed against the signal created by air passing by the pump shooters so fuel is not pulled out of the pump system. In some carburetors the discharge check is a lightweight ball at the bottom of the pump passage.

In this instance, fuel is maintained (stored) in the passage between the check and the nozzle (shooter). An anti-pullover discharge nozzle is used in such systems so air passing by the shooter will not pull fuel out of the pump passage. Because the discharge check is a lightweight ball, excess vapors in the pump can escape easily into the passage to the shooter.

This is helpful when a diaphragm-type pump is equipped with a rubber-type inlet valve because these valves seal so vapors cannot escape back to the fuel bowl. Vapors can only be purged from the pump when the pump is operated or when pressure becomes sufficient to raise the discharge ball or needle off of its seat.

Clearance around the hanging-ball-type inlet check in a diaphragm pump allows excess vapors to escape to the fuel bowl. Clearance between the pump cup and pump stem in a piston-type pump provides a vapor-escape path.

Remember, the pump's function is to replace fuel which dropped out of suspension and to compensate for fuel inertia or delay.

DESIGN FEATURES
Accelerator Pump Inlet Valves—
The synthetic-rubber valve provides an immediate pump shot with very

Accelerator-pump discharge nozzles are targeted so pump shot "breaks" against booster venturi. Photo shows how shot is pulled toward lower edge of booster venturi by air streaming into carburetor. Bubbly fuel flow from main discharge nozzle issuing from tail of booster indicates air is already mixing with fuel. Carburetor was placed on a Holley Engineering air box for this photo.

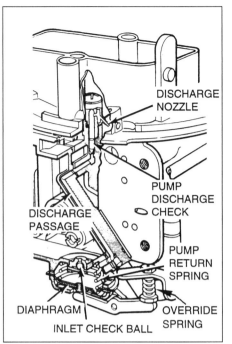

Typical pump-passage routing in carburetor with removable fuel bowls and metering blocks.

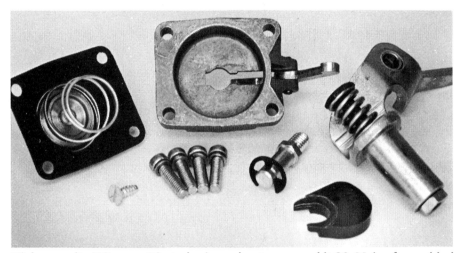

High-capacity (50cc per 10 strokes) accelerator pump kit 20-11 is often added when large secondaries are to be opened quickly. It is also helpful when there is a wild camshaft or when the carburetor is a long way from the intake ports. Pump is sometimes called a *Reo Pump* because similar ones were used on Reo trucks. Installation may require raising carburetor with a spacer so pump lever will clear manifold.

little throttle movement. This is called *tip-in capability.* Because it is normally closed, no time delay or fuel flow is required to close it. This type valve also requires the least pressure

difference to allow filling the pump. A rubber valve does not allow vapors generated in the pump to escape. And, it cannot be used with exotic fuels.

A ball-type valve is not as good for

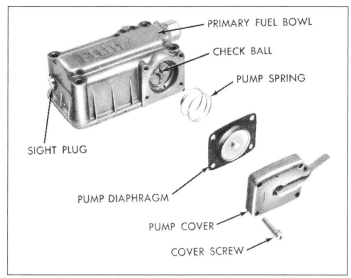

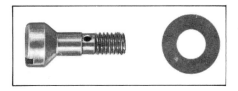

Pump discharge nozzle screw kit 26-12 includes screw with smaller shank to allow increased fuel flow to shooter. Use one of these screws with any shooter larger than 0.040 in., as it ensures the nozzle will be the limiting restriction. This is for Models 2300, 2305, 4150/4160/4180, 4165/4175 and 4500.

Diaphragm pump in removable fuel bowl showing components. Check ball used as inlet valve in this older-style bowl is no longer used. Holley currently uses rubber-umbrella inlet valves as standard, page 33. Pump is actuated by an intermediate lever operated by a cam on the throttle shaft.

Accelerator-pump discharge nozzles or shooters are marked to indicate hole diameter in thousandths of an inch. Three types shown here include anti-pullover design at left, used in 4165 and other models with lightweight pump-discharge-check valves. Two shooters at right are used with heavier check valves in the discharge passage immediately under the shooter.

tip-in because of the delay required to force the ball against its seat by part of the pump shot escaping past it. Advantages include good purging of vapors generated in the pump. Vapors escape through the clearance between the ball and the seat. Exotic fuels do not affect the steel ball.

Two Pump Sizes—All Holley high-performance carburetors (those discussed in this book) use diaphragm-type accelerator pumps because they provide very positive action. These pumps have maximum capacities of 30 cubic centimeters and 50cc. Capacity is selected to fit application requirements. Accelerator-pump capacity is measured by collecting the output from 10 full strokes of the pump. Thus, a 30cc pump delivers 3cc per pump shot at maximum stroke. Pump capacity and delivery rate are controlled by the pump cam and linkage settings.

Pump Cams—High-performance carburetors use a nylon cam on the throttle lever to operate the pump lever. A "white" cam is typically supplied on the carburetor. Cam shape affects pump action. Total lift affects the stroke and therefore the capacity available from the pump. Profile or shape of the cam controls the phasing of the pump system.

Cams are color-coded to identify each profile. See the chart on page 36 for pump capacities with various cams.

A sharp-nose cam gives a quicker pressure rise and causes strong pump action to begin immediately as the pump is activated. A gentle-ramp cam gives the opposite effect.

The pump cam is one more tool the tuner can use to make the carburetor perform well in a particular application. Several cams have been developed for specific applications. All pumps have adjustments on either the cam or the pump linkage.

Pump Discharge Nozzles (Shooters)—The pump discharges through a *shooter*. Hole size governs the rate of discharge. A larger hole

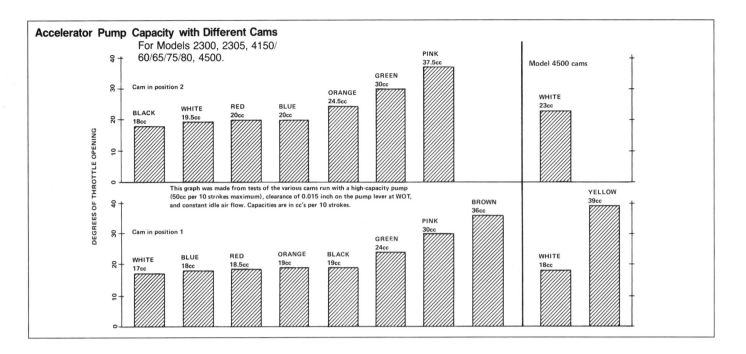

Accelerator Pump Capacity with Different Cams
For Models 2300, 2305, 4150/60/65/75/80, 4500.

This graph was made from tests of the various cams run with a high-capacity pump (50cc per 10 strokes maximum), clearance of 0.015 inch on the pump lever at WOT, and constant idle air flow. Capacities are in cc's per 10 strokes.

allows pump contents to be discharged quicker and with less pressure than a small discharge hole.

Discharge-hole size in thousandths is stamped on removable shooters. A **25** marking indicates 0.025-in. holes. Sizes range from 0.021—0.052 in.

Removable pump "shooters" or discharge nozzles are used in all Holley high-performance carburetors. Several shooter types are used and they are not interchangeable between carburetor models.

The shooter is usually aimed or targeted so fuel hits the booster venturi (if one is used). This breaks up the fuel so it is more nearly vaporized as it enters the engine. Some shooters are aimed toward the throttle plate or against the throttle bore.

Pump Override Spring—A pump override spring is used in piston and diaphragm-type pumps. It controls fuel pressure in the accelerator-pump system and pump-shot duration during wide-open throttle "punches" or "slams." When the cam or link has lifted or pushed the pump linkage to its maximum lift point, the override spring takes up the full lift travel and continues to apply pressure to operate the piston or diaphragm. Without the override spring, something would have to give or break. Because fuel is

not compressible, tremendous forces would be built up instantaneously.

Pump Selection and Timing—More details on pumps, cams, shooters, etc., are on pages 127–129.

CHOKE SYSTEM

The choke system provides the rich mixture required to start and operate a cold engine—discussed in fuel requirements, page 10. Cranking speeds for a cold engine are often around 50—75 rpm. These speeds are low compared to engine operating speeds, so little manifold vacuum is created to operate the idle system. A closed choke valve creates a vacuum below it so fuel is pulled out of both the idle and the main-metering systems during cranking.

This creates an extremely rich mixture: approximately half fuel and half air. The super-rich mixture is needed because there is not much manifold vacuum to help vaporize the fuel. Because the manifold is cold, most fuel immediately recondenses and puddles onto the manifold surfaces. And, the fuel is cold and not volatile.

Fuel from the main metering system is largely liquid because there is no air velocity to assist in atomizing it. Liquid fuel cannot be evenly distributed to the cylinders. When it arrives there, it will

not burn well. During starting, only a small portion of the fuel ever reaches the cylinders as vapor.

Once the engine starts, off-center mounting of the choke plate shaft allows airflow to open the choke slightly against a spring to start leaning out the mixture.

In the case of the automatic choke, a vacuum *qualifying diaphragm* pulls the choke valve to a preset opening once the engine starts. In some cases a temperature-modulated diaphragm varies this preset opening with ambient temperature. When the choke assumes the qualifying position it is still providing a 20 to 50% richer-than-normal operating mixture as the engine warms up and the choke "comes off" or opens.

This richer mixture is required because fuel will not vaporize well until the exhaust hot spot in the manifold warms sufficiently to ensure good vaporization. So long as the choke is "on," the engine idles quite fast: 800—1100 rpm or higher with a cold engine. The choke linkage includes a fast-idle cam to keep the engine running fast to aid in vaporizing fuel and in overcoming cold-engine friction loads.

As the engine warms up, the choke is opened fully by the bimetal spring

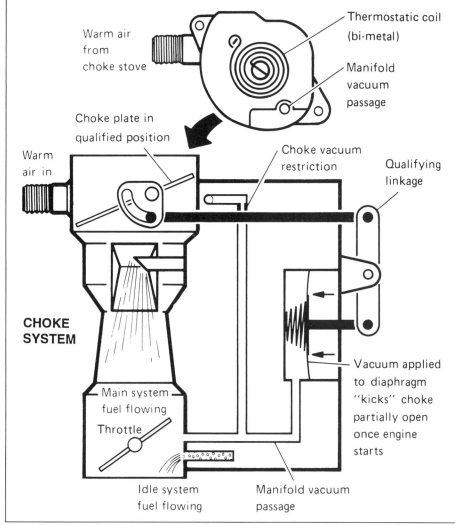

Choke system schematic.

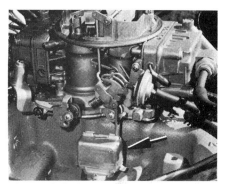

Divorced choke is actuated by remotely mounted bimetal (arrow). It is often placed in an exhaust-heated pocket in the intake manifold. Divorced chokes are subject to greater production tolerances than integral types. Holley offers Kit 45-518 to adapt 4165/75 spread-bore divorced-choke carburetors to aftermarket manifolds.

in an automatic choke, or by the operator if the choke is a manual one. This action also reduces the idle speed to a normal "curb idle."

DESIGN FEATURES

Automatic Chokes—There are two main types of automatic chokes: *integral* and *divorced*.

The integral choke is a bimetal spring inside a housing on the carburetor. A tube connects heated air from the exhaust manifold or exhaust-heat crossover to the choke. Some integral chokes use engine coolant or electricity to heat the bimetal.

An integral choke may close because there is no flow of heated air past the bimetal spring when the engine

is not running. So, the integral choke can close while the engine is still hot, even though a choke-supplied rich mixture and fast idle are not needed to start the engine and keep it running.

The divorced choke uses a bimetal spring mounted on the intake manifold or in a pocket in the exhaust-heat passage of the intake manifold. A mechanical linkage from the bimetal spring operates the choke lever on the carburetor. Divorced chokes accurately respond to engine requirements because the choke only operates when the engine is cold.

Electric Choke—Integral chokes can be operated by a bimetal spring heated by a solid-state heating unit or an electric nichrome-wire resistor (no

longer used by Holley). Either heater increases temperature similarly to that provided by the engine as it warms up. The choke operates as if the bimetal spring were sensing temperature supplied by the engine. The advantage of an electric choke is simple hook-up without plumbing or other engine attachments. Only a single wire is needed, usually from the ignition switch. Disadvantages are:

• Current is taken from the battery when power demands are high.

• A quick come-back-on activates the choke whenever the engine is turned off, even if the engine stays warm.

• Unless the engine immediately starts when the ignition is turned on, the choke opens, even though the engine is still cold. This problem can be fixed by powering the choke from the alternator circuit so the choke only gets power when the engine is operating. Switch 12-810 can be used for this with wires connected to the C and NO terminals.

• The choke can come off while the engine is still cold.

Choke Index—Integral automatic chokes have index marks. The factory setting causes the choke to close (on a new carburetor) when the choke bimetal is at approximately 70F (21C). Tapping the carburetor lightly overcomes any shaft friction so the choke seeks the position being set by the bimetal.

If less choke is desired at this temperature, move the choke index one mark. An arrow on the housing shows the direction to move the index to change the choke operating characteristic (eiher rich or lean). A choke mixture change rarely requires moving more than one index-mark from the factory setting.

Unloader—If the engine does not start quickly, extra fuel fed by the

Electric choke caps. Top one (obsolete) used nichrome-wire resistance heating element to warm bimetal and open choke. Unit at bottom (used since mid-'80s) uses a solid-state heating device powered from vehicle's 12-volt source.

WHAT'S A BIMETAL?

The bimetal is two different metals bonded into a strip and formed into a coil. Because the metals have different thermal-expansion characteristics, the coil unwraps when warmed and wraps up again when cooled. The outer or "free" end of the coil attached to the choke linkage holds the choke closed. Or, it loads the plate so the plate closes when the throttle is opened. When the bimetal is warmed, the choke opens. Warming the bimetal is accomplished with exhaust-gas-warmed air, jacket water or an electric heating element. Bimetal temperature-response characteristics are built-in according to the metals used. Most choke bimetals wrap up (choke just closed) at 65F to 70F (18C–21C).

Examples from left to right: Manual choke (arrows indicate cable-housing and lever clamps), integral choke with bimetal and housing removed to show built-in vacuum qualifier piston (arrow), divorced choke with vacuum-qualifier diaphragm (arrow) attached to carburetor body. .

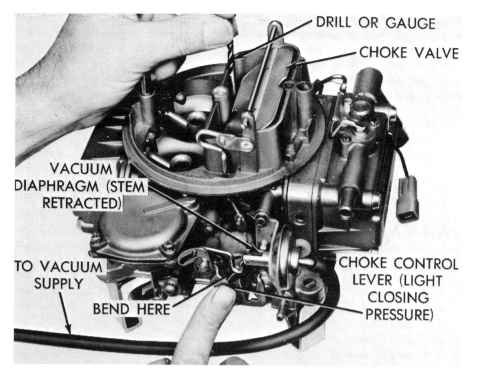

DRILL OR GAUGE
CHOKE VALVE
VACUUM DIAPHRAGM (STEM RETRACTED)
TO VACUUM SUPPLY
BEND HERE
CHOKE CONTROL LEVER (LIGHT CLOSING PRESSURE)

Vacuum qualification setting is detailed in repair-kit instructions and tune-up specifications. In some cases check is made on down side of choke plate—as here—in others on the up side.

Choke index can serve as guide when changing choke settings. Limit such changes to one mark RICH or LEAN as noted on choke housing. Three screws must be loosened to allow adjustment.

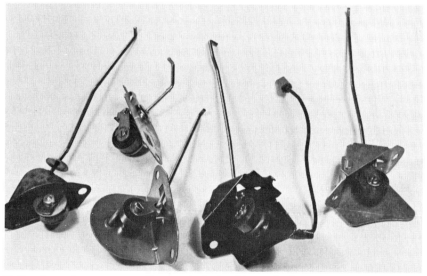

Collection of remote or divorced choke bimetals from GM and Chrysler vehicles. Some have been made with Calrod heaters next to bimetal to program choke come-off to meet emission standards. Pre-1972 units sometimes had adjustable bimetals to allow changing preload.

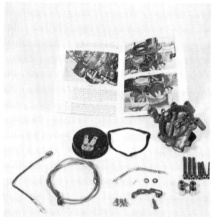

Electric-choke conversion kit 45-224 uses external vacuum supply. Fits 0-4412, 0-4776-2 through 0-4781-2, 0-6299, 0-6708, and 6709.

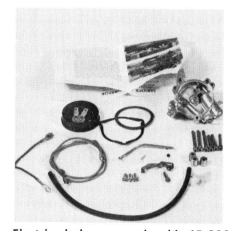

Electric-choke conversion kit 45-223 converts carburetors with an internal vacuum supply. Also works with 0-1850, 0-3310, 0-6425 and 0-7448.

choke to the engine will create an overrich mixture which cannot be burned. This must be cleared (or purged) from the manifold with air. Crank the engine while holding the throttle wide open (push accelerator pedal to the floorboard) and open the choke so additional fuel is not drawn in. With an automatic choke, opening the throttle wide causes a tang on the throttle lever to contact the fast-idle cam. This unloads (opens) the choke plate sufficiently to allow clearing excess fuel from the manifold.

SECONDARY SYSTEM

For many years U.S. carburetors were single-stage carburetors. V-8 engines were equipped with a two-barrel. It was merely a single casting containing two one-barrel carburetors side by side. One was used for each level of the traditional two-level cross-H manifold. So the usual two-barrel is a *single-stage* carburetor.

The late '40s saw an increased emphasis on vehicle performance. Because more airflow means more power, single-stage carburetors became bigger and better—with accompanying driveability problems. The large venturis required higher rpm or greater airflow to start main-system flow. Under some conditions, the idle system could be made to cover up the late entry of the main system. But, because the idle system is controlled by manifold vacuum, a great deficiency was felt at low manifold vacuum and low airflow. For example, when driving in a high-load, low-rpm condition.

In addition, the low venturi velocities at low rpm resulted in poor fuel vaporization and poor fuel/air mixing. Cylinder-to-cylinder distribution problems and erratic operation of the engines at lower airflows were common.

Simply stated, the metering range of the single- or two-barrel carburetor was too narrow to satisfy all driving requirements. The answer was obvious: use a *staged* carburetor to stretch the metering range. This allowed using venturis small enough to get the main systems flowing at low rpm. This provided good vaporization and the required capacity for high-rpm operation when it was needed.

The first staged two-barrel on a U.S. car was on the 1970 Pinto. Such systems had long been used on European and Japanese cars. Their costs were outweighed by the need to get maximum economy and maximum performance from small-displacement engines.

In the early '50s, a whole host of four-barrel carburetors were introduced for V-8s. The primary side of these carburetors was just like the older single stage carburetors. Primary venturi size was smaller than it had been on the single-stagers, returning all of the benefits provided by small venturis. These include early start of the main system, good vaporization and fuel/air mixing, and good distribution. Primary carburetor barrels were used for cruising loads and light accelerations encountered in normal traffic. Flexibility of operation and economy were regained.

Two secondary carburetor barrels were coupled to the primary carburetor barrels. These operated when maximum airflow was required for more power. The carburetor's metering range

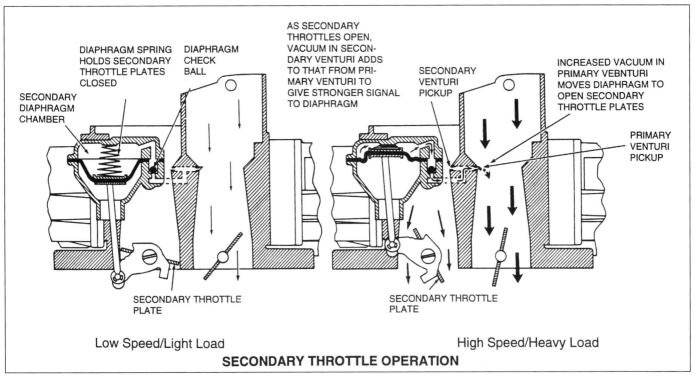

SECONDARY THROTTLE OPERATION

Schematic of diaphragm-operated secondary throttle. Signal from primary venturi is augmented by an increasing signal from secondary venturi as secondary throttle starts opening. Closing primary throttle closes secondary by mechanical over-ride linkage. As signal from venturi ports dies (pressure closer to atmospheric) check valve blows off seat to remove vacuum from diaphragm chamber. Secondary opening rate is controlled by spring behind diaphragm and size of restriction through which vacuum is applied.

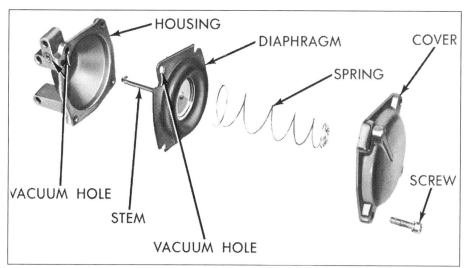

HOUSING

DIAPHRAGM

COVER

SPRING

SCREW

VACUUM HOLE

STEM

VACUUM HOLE

Exploded view of secondary-throttle diaphragm. Changing spring varies varies secondary-throttle opening point. A quick-change cover in kit 20-59 allows changing springs by removing only two screws instead of six, greatly simplifying tuning.

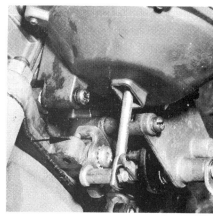

Diaphragm-operated secondary opens as engine needs added airflow. This type of secondary operation is very "forgiving" in terms of size selection. A too-large carburetor (airflow capacity) can often be used without spoiling low-end performance and drivability. Arrow indicates secondary-throttle stop.

Mechanically actuated secondary throttles operate progressively as shown here. Primary opens 40° or so before secondary opening begins. Linkage opens secondaries fully as primary throttles reach full-open.

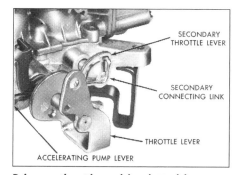

SECONDARY THROTTLE LEVER

SECONDARY CONNECTING LINK

THROTTLE LEVER

ACCELERATING PUMP LEVER

Primary throttle rod in slotted lever on secondary shaft will not allow secondary-throttle opening beyond an equivalent primary-throttle opening. Even though primaries may be opened wide, secondaries will not start to open until there is sufficient airflow through the primary venturis to cause the diaphragm to actuate the secondary throttles.

was essentially doubled. Consequently, good part-throttle operation was combined with relatively unrestricted flow for maximum-power conditions.

The secondary side is simply another carburetor. It usually opens later than the primary. Secondaries always have their own main-metering system. Most have an idle system because this gives better mixture distribution and idle stability. The result

is improved emissions performance because leaner idle settings can be used. The idle system also keeps the fuel level from rising in the secondary bowl when the needle and seat leak slightly or open due to float bouncing. A secondary idle system also ensures fresh fuel in the secondary bowl.

In 1967 Holley began providing some carburetors with accelerator pumps for both primary and sec-

ondary barrels. Some carburetors have power systems in the secondary side.

Secondary operation can be activated in several ways:

- Mechanical only.
- Diaphragm.
- Mechanical with velocity valve.
- Mechanical with air valve.

Holley carburetors use the first two methods.

Mechanically operated secondaries are simple in operation. Secondary throttles are opened by a direct link from the primaries, usually progressively. Secondary opening is delayed until primaries are approximately 40% open. Secondary-throttle closing is positive because of the return spring, because throttles are offset on their shaft so airflow aids closing and because a return link from the primary throttle pulls them closed.

Holley pioneered the use of separate accelerator pumps on mechanically actuated secondaries, now considered an essential feature for high-performance applications. The secondary pump ensures adequate fuel to carry the engine through any period when the driver opens the throttles quickly.

This is important when they are opened at an rpm too low to generate

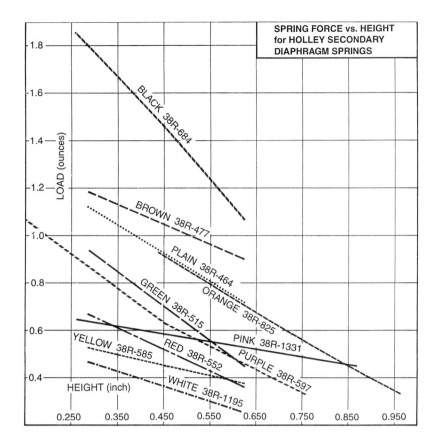

SPRING FORCE vs. HEIGHT
for HOLLEY SECONDARY
DIAPHRAGM SPRINGS

sufficient signal for main system flow on both the primary and secondary sides of the carburetor. The pumps actually provide mechanical fuel injection during the time the signal is being reestablished.

Diaphragm-operated secondaries have also been highly refined by Holley. These carburetors use a diaphragm to open the secondary throttles. Vacuum for the diaphragm is obtained from one of the primary venturis. Part of this signal is usually bled off through an opening or bleed into one of the secondary venturis. As the engine is operated at increasingly higher rpm, velocity through the primary venturi creates a vacuum signal.

The amount of secondary opening depends initially on airflow through the primary barrel. As the secondary throttles begin to open and flow is established in the secondary barrel, the vacuum signal from the primary is augmented by vacuum from one of the secondary barrels through the previously mentioned bleed opening.

Secondary throttles are fully open at maximum speed. But, if the carburetor is too large for the engine, the diaphragm automatically sizes the carburetor. The diaphragm partially opens the secondary throttles to supply only the needed amount of mixture. Secondaries remain closed when the primary throttles are opened wide at low rpm. This eliminates tiny bogs and allows using the carburetor with a wide range of engine displacements, gear ratios, car weights and so forth.

A ball check in the vacuum passage to the diaphragm applies vacuum at a controlled rate through a bleed (a groove in the check-ball seat). When rpm is reduced so secondary airflow is not needed, the ball check instantly releases vacuum from the diaphragm so the throttles are closed by the return spring behind the diaphragm and by airflow acting against the throttles which are offset on their shaft. The mechanical link positively closes the secondaries when the primaries are closed.

The diaphragm-operated secondary has an especially low opening effort or pedal "feel" because the throttle linkage only has to open the primaries and only actuates one accelerator pump. Secondaries are opened by the diaphragm without any assistance from the driver.

FEEDBACK CARBURETION

In general, gasoline engines produce three harmful emissions: hydrocarbons (HC), carbon monoxide (CO) and oxides of nitrogen (NO_x). The amount of each emitted by a vehicle under controlled test conditions is limited by law.

Emission standards have become more and more difficult to achieve over the years, and things are further complicated by the Corporate Average Fuel Economy (CAFE) standards. Changes made to satisfy one set of standards may negatively affect the other.

Catalytic Converters—In 1975 the exhaust catalyst was introduced as a method of handling HC and CO. The original catalytic converters were oxidizing catalysts—they encourage both compounds to take on more oxygen. Carbon monoxide is changed to carbon dioxide (CO_2), which is harmless, and hydrocarbons are converted to water (H_2O) and carbon dioxide.

Oxides of nitrogen (NO_x) were lowered by engine modifications that lowered combustion temperatures. Most of the improvement was accomplished through lower compression and exhaust-gas recirculation (EGR). EGR dilutes the intake charge with exhaust gas.

Closed-Loop Control—Further reductions of NO_x required more advanced technology, which brings us to the *feedback* or *closed-loop* system. With this system the fuel/air mixture is controlled by engine demand conditions and not just by the air flowing through the carburetor. If the concentrations of HC, CO

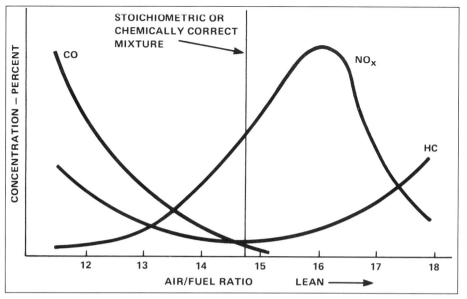

Approximate relationship between CO, HC, and NO_x as air/fuel ratio changes. Closed-loop control is designed to keep the ratio close to stoichiometric, 14.7:1.

Bosch Lambda sensor (oxygen sensor) is mounted in exhaust system upstream of catalytic converter. Its 0—1 volt output is sent to ECU at right.

cally correct or *stoichiometric* air/fuel ratio, approximately 14.7:1.

Maintaining the emissions balance is the major reason for the closed-loop or feedback system. A system of engine sensors supplies data to a controlling microprocessor that manipulates a mixture adjuster in the carburetor.

The fuel/air ratio is monitored from an exhaust-gas sample measured by an *oxygen sensor* in the exhaust system. This typically sends a voltage output for interpretation by an *electronic control unit* (ECU, a microprocessor). It, in turn, alters the fuel/air mixture in the carburetor with a *solenoid-controlled valve* (also called a *duty-cycle solenoid*) to maintain the emissions balance.

If the air/fuel mixture is kept close to 14.7:1, the three-way converter's efficiency is high for all three exhaust products. A lean mixture lessens the converter's efficient handling of NO_x. A rich mixture causes an efficiency drop for handling CO and HC.

Oxygen Sensor—In addition to the three pollutants mentioned, oxygen concentration also varies with fuel/air ratio in a predictable manner. Consequently, the amount of oxygen in the exhaust can be used as a guide to adjust the fuel/air mixture.

The oxygen sensor is placed in the exhaust upstream of the catalyst. The sensor puts out a low-voltage signal to the ECU. As oxygen content increases—lean mixture—the voltage output drops. As it decreases—rich mixture—the voltage output increases. These output signals are used by the ECU to adjust the fuel/air ratio accordingly.

The sensor's total output ranges from 0—1000 millivolts (mv) (1.0 volt = 1000mv). A lean mixture would typically be indicated by a reading of 100—200mv. Rich mixtures are found at 700—900mv. A defective sensor doesn't change voltage as the fuel/air mixture is altered, or its voltage output changes very slowly as oxygen content of the exhaust increases or decreases.

and NO_x can be held in proper balance they can all be treated by a single catalyst called a *three-way catalyst*. It uses palladium and rhodium as agents to reduce NO_x to harmless nitrogen and oxygen.

What happens in practice is straightforward. Oxygen released from NO_x combines with HC and and CO. No_x becomes free nitrogen (N_2)—an inert gas under normal conditions—and HC and CO become water and carbon dioxide. Maintaining the emissions balance is the major reason for the closed-loop or feedback system. It determines fuel/air ratio from an exhaust-gas sample and changes fuel metering to maintain the balance.

The design challenge is in maintaining the correct balance. The ideal combination occurs near the chemi-

Electronic Feedback Carburetor

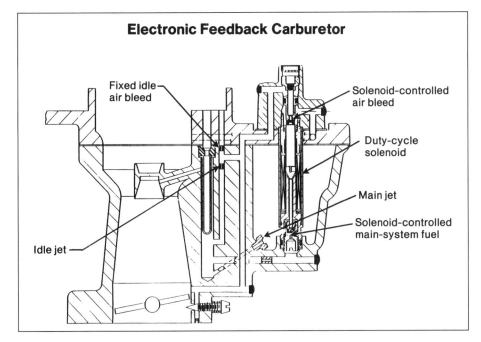

Fixed idle air bleed

Solenoid-controlled air bleed

Duty-cycle solenoid

Main jet

Solenoid-controlled main-system fuel

Idle jet

Model 6510 for 1981 Chevette. Duty-cycle solenoid is connected to electrical lead at left. Throttle-position sensor (arrow) supplies input to ECU. Although not covered in detail in this book, this carburetor is used as an example to explain the concept.

A correctly operating sensor will send minute fluctuating voltages, i.e, 910, 920, 900, 930mv as it monitors exhaust oxygen content. A sensor's effective life is usually rated at 30,000 miles.

Unless a *heated* oxygen sensor is used that heats to operating temperature almost instantly as the ignition is turned on) the oxygen sensor must be at 600F (315C) to supply reliable data to the ECU for adjusting the mixture. Consequently, during cold starts with an *unheated* oxygen sensor, closed-loop control reverts to open-loop control of the mixture.

Open-Loop Control—During warmup and heavy throttle applications, richer mixtures are required. These richer driving cycles are called *open-loop* because the oxygen sensor in the exhaust stream is not supplying data to the ECU for controlling the fuel/air mixture.

Mixture control at cold starting is regulated by programmed instructions stored in the ECU's memory. When the oxygen sensor operating temperature is reached, 600F (315C), and coolant temperature has risen, the ECU accepts data from the oxygen sensor and closed-loop control takes over.

Two sensors usually supply the necessary input to the ECU for WOT mixture control: A *vacuum sensor* (manifold pressure) and a *throttle-position sensor*. When these indicate heavy throttle application, the system goes back into open-loop mode and a rich mixture is supplied. When this requirement stops, the ECU switches back to closed-loop control and relies on the oxygen-sensor signal input to alter the solenoid-controlled valve's activity.

Vacuum Feedback System—On the vacuum feedback system, using the Holley 6500 as an example, the low-voltage signal from the oxygen sensor is fed through an amplifier to a switch. This switch controls a valve that channels manifold vacuum to a special diaphragm on the carburetor. The diaphragm mechanically controls a bleed to a revised power valve.

The vacuum feedback system (Holley 6500) was introduced on the 1978 Ford 2.3-liter engines for California. This carburetor is not covered in detail in this book, but is used as an example to explain the concept.

Electronic Feedback System—Another type of feedback carburetor was introduced on the 1980 Chevette (Holley 6510) and Omni/Horizon (Holley 6520). The principle was subsequently extended to many other models.

In this feedback system, the signal voltage originates at the oxygen sensor, and the ECU generates a 10-Hertz (10 cycles per second) signal that is sent to a solenoid-controlled valve in the carburetor.

This duty-cycle solenoid opens and closes at the frequency sent by the ECU. The percentage of time

Vacuum Feedback Carburetor

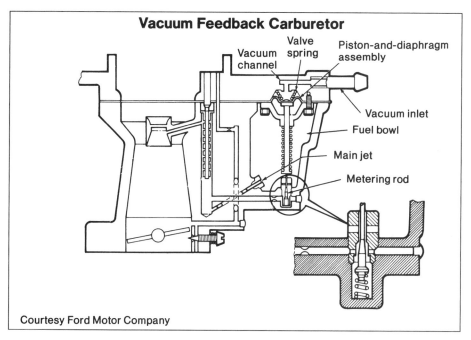

Labels: Valve spring, Vacuum channel, Piston-and-diaphragm assembly, Vacuum inlet, Fuel bowl, Main jet, Metering rod

Courtesy Ford Motor Company

Duty-cycle solenoid removed from carburetor. Actual fuel-control valve is at bottom. Two O-rings (arrows) keep fuel from flowing around valve.

TUNE WHILE YOU DRIVE

You can tune your Holley *while driving* with the Quarter Mile Dial and Mile Dial. A duty-cycle solenoid in the fuel bowl is electronically controlled by a signal generator mounted inside the car. You can adjust the solenoid duty cycle and control the fuel flow. With the Quarter Mile Dial or performance version, fuel flows into a passage connecting the two main-jet wells for that side of the carburetor. So, all metering above idle is affected by the solenoid adjustment you dial in. You can electronically adjust in a range of about five jet sizes. Every switch position is one-half jet size or about 2%.

Quarter Mile Dial conversions have a solenoid-equipped bowl for both primary and secondary on carburetors with mechanical secondaries; on the primary side only where they are vacuum-operated (diaphragm) secondaries.

On the other hand, the Mile Dial has a single solenoid-equipped bowl on the primary side. The solenoid orifice is slightly smaller than that in the Quarter Mile Dial. Fuel flows from the solenoid feeds the power-valve circuit and flow is limited by the size of the power-valve restrictions. Once the valve opens, the Mile Dial has no effect on the mixture during part-throttle operation when the power valve isn't operating. As an example, cruising on the highway.

the valve remains closed (the duty cycle) during each cycle varies, depending on the signal from the ECU.

For example, under certain conditions the valve might be open 6 milliseconds (ms), closed 4ms, open 6ms, closed 4ms and so on. If the oxygen sensor indicates this is too rich a mixture, the valve's timing could be altered by the ECU to open

only 4ms and close 6ms. Either way, each total cycle, (open *and* closed together) is 10ms.

On some carburetors this valve only operates on the main system by opening and closing a restriction to supplement the main jet. Other carburetors have valves that also open and close an auxiliary air bleed affecting the idle mixture as well.

Mixture is controlled by the length of time the valve is held open or closed. The ECU's output to the valve is constantly altered as it monitors the different sensors: oxygen, throttle-position, manifold-pressure, coolant-temperature and so on that indicate engine demand..

Driver Control of Mixture—The technology used in the closed-loop system can be used to improve fuel economy. With this in mind, Holley introduced a driver-controlled *open-loop* system, the Quarter Mile Dial. Although the Quarter Mile Dial allows controlling fuel economy, its primary use is for dialing in the best power under prevailing conditions. This is accomplished without removing the fuel bowl or changing main jets. You can use a stopwatch to time acceleration, but sophisticated drag-strip timing is the ultimate. Some racing associations have ruled these systems as legal bolt-on equipment.

Quarter Mile Dial systems are a useful tool for anyone doing dynamometer test work and setting best fuel at every load point. The system provides a 20% fuel-adjustment range. Changing the main jets moves the "range of authority" up or down as desired.

In this system, the feedback control is replaced by a driver-controlled switch/mixture-control box which is a signal generator.

The solenoid-controlled air bleed is eliminated, so the duty-cycle solenoid affects only fuel flow into the main well.

The solenoid is controlled from an 11-position switch that changes the duty cycle (the percentage of time the solenoid is closed) from maximum rich to maximum lean. You can electronically adjust in a range of about five jet sizes.

You can set the system for maximum economy on the street or tune for the best F/A ratio for maximum power. This can be done for any combination of engine components being used and existing atmospheric conditions. Holley markets this equipment as Quarter Mile Dial.

Let's look first at the equipment to adapt to the popular Model 2300 and 4150/4160 carburetor families. Extra fuel is introduced downstream of the power-valve restriction to control both part-throttle and WOT mixtures.

Kits or complete systems are available. Kits contain the mixture-control box, one or two metering blocks, one or two fuel bowls, gaskets and a wiring harness. Systems consists of a complete carburetor equipped with a duty-cycle solenoid, plus the mixture-control box and wiring harness.

If you already have one of the adaptable carburetors, you can use a kit. If not, buy a complete system.

Kits and systems for various applications are listed in the Holley Performance Parts Catalog.

Mile Dial Carburetors—This different group of carburetors and kits provides mixture control for specific passenger cars and trucks. While the control system is similar to that used on the previously discussed Quarter Mile Dial, these carburetors channel fuel from the duty-cycle solenoid to the power-valve circuit instead of to the main well. Kits and systems for various applications are listed in the *Holley Performance Parts Catalog*.

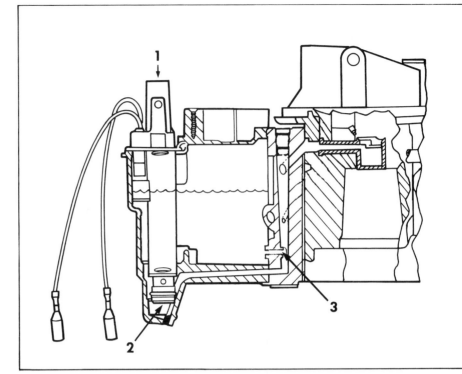

(Above) Quarter Mile Dial schematic: (1) duty-cycle solenoid, (2) auxilary jet, and (3) screw-in jet. Solenoid is adjusted by driver for optimum fuel flow through main system under different operating conditions.

(Below) Quarter Mile Dial for 4150. Some systems have a solenoid just in primary bowl.

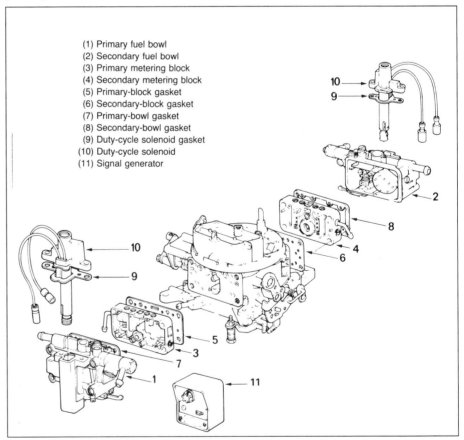

(1) Primary fuel bowl
(2) Secondary fuel bowl
(3) Primary metering block
(4) Secondary metering block
(5) Primary-block gasket
(6) Secondary-block gasket
(7) Primary-bowl gasket
(8) Secondary-bowl gasket
(9) Duty-cycle solenoid gasket
(10) Duty-cycle solenoid
(11) Signal generator

High-Performance Carburetors **3**

Part 0-4412 Model 2300 has center-pivot race bowl with adjustable needle/seat and sight plug for setting fuel level externally. High-capacity accelerator pump (50cc per 10 strokes) is used. Power valve is backfire-protected. Hand choke linkage is standard. The carburetor base fits Ford 2-barrel manifolds. An adapter is required for use on other engines. This 500 cfm performance carburetor is for modified engines; it is not suitable for street use on stock 289- or 302-cid engines.

2300 flange with backfire-protection valve (arrow). This is carburetor body side of flange. Spring-loaded ball is pulled off its seat by manifold vacuum so power valve can operate. It slams shut when a backfire occurs, protecting the power-valve diaphragm from being ruptured. This one was factory-installed, but you can add this protection to any 2300, 4150, 4160, 4165, 4175 or 4180 carburetor, see page 55.

MODEL 2300

Model 2300 carburetors have been used on vehicles ranging from Jeeps to muscle cars. They have been reliably used for years on Ford passenger cars and trucks, Chevrolet trucks, International Harvester, American Motors, White Motors, Reo, and Willys Jeep.

Venturi sizes have ranged from 1- to 1-7/16-inch diameter with airflows from 210 to 600 cfm. The largest 2300 currently made flows 500 cfm.

Chevrolet musclecars and Dodge/Plymouth "Six-Pack" triple manifolds used diaphragm-operated 2300's as the end or outboard units.

These have no chokes or accelerator pumps and use metering plates like the Model 4160. The center 2300 has a choke, accelerator pump, and a metering block with screw-in main jets and a power valve. These are available as replacements for restoration purposes.

Description of Operation

Because the 2300 is literally the front half of a 4150/4160 carburetor, operation is just like that described for the Model 4150 on page 53.

Some 2300's have a "reverse" idle adjustment as described and illustrated in the 4150 section. A metering-block label identifies this idle system.

In 1992 Holley began including power valve blowout protection in the most-popular Model 2300s. Older carburetors can be equipped for this protection, see page 55.

Hose connections in the carburetor base are for timed and vacuum spark advance and for a PCV valve. The timed port can also be used to purge charcoal canisters. An SAE 1-1/2-inch spread-bore pattern (3-1/2-inch x 5-1/2-inch hole spacing) fits Ford two-barrel 289- through 390-cid engines.

Non-Ford engines require an adapter to mate the bolt pattern and 1-11/16-inch throttle bores to the manifold flange. Adapter 17-19 fits the SAE

Model 0-3310-4 is identified by stamped number on vertical portion of airhorn casting. -4 on this particular carburetor identifies the changeover to all-metal parts. Bottom number is build date. First three numbers are day of year—076, or 76th day—March 16th. Last number, 2, indicates 1992. Of course, the year can be off by one or more decades. So when figuring out how old a carburetor is, it is helpful to know identifying features and when they were introduced. Complete bowl vent as part of airhorn casting identifies this as a current model.

HOLLEY NUMBERS

A complete Holley assembly—such as a carburetor—can be referred to by three different numbers: model number, engineering part or list number, and sales number.

Model Number—Model numbers have meaning. They describe the class, type and general features of products in that group. Model 4150 describes a basic four-barrel. Many different calibrations and individual features exist on various part/list numbers within the same model grouping.

The first digit usually designates the number of barrels (throttle bores). Example: Models 2010 and 2300 start with 2 and are two-barrels. The other digits indicate variations making one different from the other. 4150 and 4160 are similar four-barrels with secondary-metering and throttle-actuation differences. 4010 and 4011 are four-barrels.

As with any numbering system, there are exceptions. For instance, models 1901 and 5200 (not discussed in this book) are two-barrels. Numbers beginning with 6 are usually "closed-loop" or feedback carburetors. Here too, these are exceptions.

Engineering Part Number—assigned to carburetor or other assembly in strict numerical sequence without regard for any other factor. If a two-barrel carburetor, a fuel pump and a four-barrel carburetor were initiated in that succession, their part numbers will follow that order. These numbers are then assigned to their model groupings. Engineering numbers were originally set up with an R- prefix for the assembly number and on group numbers of components. Example: The Part No. 0-4776 double-pumper high-performance carburetor has also been known as List R-4776-AAA, List R-4776-1AAA, and R4776.

The prefix letter may be changed to assist identification, such as P- for fuel pumps. Performance carburetors now use a 0- prefix, but may also be listed with an R- prefix.

The part number is stamped on the carburetor. On the 2300, 4150/4160, 4165/4175, 2010, 4010/4011 it is on the airhorn extension supporting the choke shaft. On 4500s it is stamped on a flat part of the airhorn.

Sales Number—This number is assigned in numerical sequence for popular replacement carburetors and service items such as needle/seat, pump piston, pump diaphragm, repair kits, gasket kits, and so forth.

Sales numbers include a prefix number, a dash and a suffix number. Prefix 1- identifies carburetors, with the suffix identifying an individual carburetor. 1-108 completely identifies a particular replacement carburetor. The number is used for merchandising and computer data processing. It does not appear on the carburetor, even though it may be used on the carton and in price sheets. Performance carburetors use a 0-prefix.

Carburetor-service components are also identified by product categories or groups. Prefix numbers are assigned to screws, gaskets, throttle plates, etc. Suffix numbers identify the particular parts.

1-1/2-inch 2V pattern used on some late Chevrolets. Chrysler products (318 cid) and some early Chevrolets have a SAE 1-1/4-inch 2V pattern which work with Adapter 17-3.

The 2300s can be equipped for driver control of the fuel mixture.

Quarter Mile Dial kits 34-103 and 34-104 allow electronic adjustment within a range of about 5 jet sizes at wide-open throttle. Details of the kits are on page 45.

List 0-7448, 350 cfm—One of the most-often specified 2300s, this car-

buretor has two 1-3/16-inch venturis fed by 61 main jets. The power valve is a 8.5-in.Hg unit. A 30 cc per 10 strokes accelerator pump is used. The center-pivot-float race-type bowl has a sight plug and externally adjustable needle-and-seat assembly.

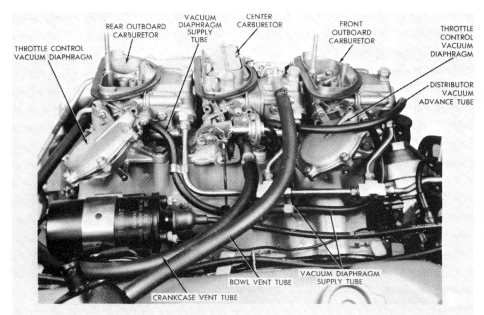

REAR OUTBOARD CARBURETOR

THROTTLE CONTROL VACUUM DIAPHRAGM

VACUUM DIAPHRAGM SUPPLY TUBE

CENTER CARBURETOR

FRONT OUTBOARD CARBURETOR

THROTTLE CONTROL VACUUM DIAPHRAGM

DISTRIBUTOR VACUUM ADVANCE TUBE

BOWL VENT TUBE

VACUUM DIAPHRAGM SUPPLY TUBE

CRANKCASE VENT TUBE

Diaphragm-operated outboard carburetors on 3 x 2 Chrysler manifolds are opened by vacuum signal from center carburetor as engine requires extra airflow. Only center carburetor has accelerator pump. Holley offers 500 cfm outboard carburetors with accelerator pumps for use with progressive mechanical linkage. Manual linkage that operates all three carburetors simultaneously is not recommended.

Two Model 2305 Primary Plus carburetors are the 500 cfm 0-80095 at left and the 0-80120 350 cfm. 500 cfm is identifiable by additional struts cast under the airhorn extension. Features annular booster on the primary barrel, side-pivot fuel bowl with swivel (banjo) input and staged progressive linkage.

Staged progressive linkage operation is shown here. Note how primary leads secondary opening. This greatly improves drivability.

There is a manual choke. Power-valve blowout protection was added in 1992.

The universal linkage fits the carburetor to GM and Ford applications, including automatic-transmission kickdown, but not Fords with automatic-overdrive transmissions. Part 20-7 adapts the 0-7448 to Chrysler applications.

List 0-4412, 500 cfm—The 0-4412 Model 2300 introduced in 1969 became an instant success among engine builders and tuners preparing engines to race in classes limited to one two-barrel carburetor. Two 1-3/8-inch venturis are fed by 73 main jets. The power valve is a 5.0-in.Hg unit. A high-capacity (50 cc per 10 strokes) accelerator pump is used. The center-pivot-float race-type bowl has a sight plug and externally adjustable needle-and-seat assembly.

There is a manual choke and power-valve blowout protection is included ass of 1992.

The universal linkage fits the carburetor to GM and Ford applications, including automatic-transmission kickdown, but not Fords with auto-

matic-overdrive transmissions. Part 20-7 adapts the 0-7448 to Chrysler applications.

List 0-9647 Methanol Carburetor—Racers will be interested in this Model 2300 that is calibrated for use with methanol fuel.

MODEL 2305

As interest developed in trying to obtain performance from 4-cylinder engines, it became apparent that the 270 CFM provided by the original equipment Holley-Weber Model 5200 (Pinto, Vega, etc.) was not enough air-

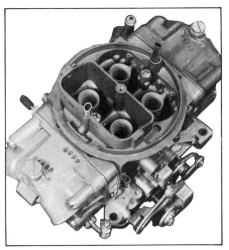

Model 4150 0-9381 is competition unit with annular-discharge booster venturis. This 830 cfm carb has dual-inlet center-pivot fuel bowls, double accelerator pumps and adjustable secondary idle mixture screws.

flow capacity. So Holley introduced the Model 2305 in 350 and 500 cfm versions. From the throttle body up, it is essentially the same as the Model 2300.

The 2305 is a *staged* carburetor. Mechanical linkage starts opening the secondary as the primary reaches approximately 40° of angular travel. Primary and secondary bores have individual throttle shafts positioned 90° from the normal 2300's single shaft. Throttle linkage is opposite the fuel bowl.

Throttle-bores are 1-11/16" (43mm) in both versions. Airflow is controlled by venturi size: 1-3/16 in. in the 350 cfm 0-80120; 1-3/8 in. in the 500 cfm 0-80095. Annular-discharge venturis are featured. Throttle-bore and stud spacing are exactly the same as the Model 2300.

The accelerator pump is actuated by the primary shaft. Pump nozzles discharge simultaneously into the primary and secondary sides.

Side-hung fuel bowls have sight plugs and externally adjustable needle-and seat assembly. Bowls are equipped with a 5/16-in. fuel-hose fitting. A center-hung float bowl, kit 34-15, is also available.

Manifold adapters are available for Ford, Chrysler, Datsun, Pontiac and Toyota 4-cylinder engines.

A big advantage in being part of the Model 2300 family is that most service parts are interchangeable and readily available: power valve, main jets, pump nozzles, etc. Even the Quarter-Mile Dial performance kits will fit.

These carburetors are *not bolt-on replacements* by any stretch of the imagination. If you want to use one, consult the Holley Standard Replacement Catalogs. These carburetors are for enthusiasts and racers whose engines need more airflow. Expect to do a little "cobbling" or improvising when installing a 2305.

List 0-80120, 350 cfm—this carburetor is intended primarily for street performance. It is equipped with a manual choke. Kit 45-385 converts it to an electric choke and 45-836 is a hot-air choke conversion.

List 0-80095, 500 cfm—This competition carburetor has no choke.

FOUR-BARREL CARBURETORS

Current Holley four-barrel carburetors fall into five basic families: 4150, 4160 and 4180 are described here. The now-obsolete 3150 and 3160 three-barrels have similar construction and operation. 4165/4175, 4500, 2010, 4010 and 4011 are separately described.

MODEL 4150

Holley 4150 carburetors were first introduced in 1957 on Ford 312-cid-engine automobiles. And, while we like to think of these units as strictly high-performance devices, they have seen widespread use on normal passenger cars and trucks. Some even have governors! These carburetors have some interesting design features.

Center-pivot fuel bowls as used in current Holley Carburetors were developed for Chevrolet applications. Conversion kits are available to replace side-hung bowls with the center-pivot type: Kit 34-2 converts 4150/60 carburetors.

The original center-pivot float bowls were a "snout-type," first used on Ford and Chrysler NASCAR racing engines in 1964. They were also used on Shelby Mustangs and Cobras. Ford performance and restoration enthusiasts can buy the Le-Mans Fuel-Bowl Conversion, Kit 34-14. These bowls do not work with double-pumpers.

• Externally adjustable floats with sight plugs for setting fuel level, first used on 1957 Ford Model 4150s. Available on both side-hung and center-pivot float bowls.

• Bowl vent plastic "whistles," first used in carburetors supplied for 1966 Fords.

• Double accelerator pumps, first used on List 4296 4150s for 1967 Chevrolet big-block applications.

• Mechanical secondary linkage, first used for Chrysler in 1959 (List 1970).

• Center-mounted accelerator pump discharge nozzle, designed for plenum-type tunnel-ram manifolds and first used in 1967 (List 4224).

• Annular-discharge boosters, first used on Model 4180 on 1979 Ford trucks.

• Carburetor bases held to main body with six screws instead of eight starting in 1982.

Model 4180 is a much-modified Model 4160. Significant changes in the main and idle systems improve fuel metering, calibration and control. Sealed idle-mixture screws are in the throttle body instead of the metering block. Primary boosters are annular-discharge type. Used as original equipment on 351- and 460-cid Ford trucks and on high-output 302-cid engines in Mustangs and Capris.

4160 or 4180 die-cast secondary metering plate. Fuel and air channels are cast-in. Drilled holes in bottom of casting provide main metering. As shown here, metering plate may be used with a flat steel plate 1034-1993 plus an extra gasket.

Top side of this Model 4160 3310-4 four-barrel throttle body has the power-valve blowout protection factory-installed (arrow). This protection can be added to older carburetors, see page 55 for details.

• Hex-head screws used to attach fuel bowls beginning in 1987.

• Pump-inlet valves. Rubber "umbrella" accelerator-pump inlet valves became standard in all performance carburetors in 1990.

• Power-valve blowout protection for 4150/4160 introduced in 1992. Older carburetors can be retrofitted with this important feature, page 55.

Numerous 4150 carburetors have been used as original equipment by the Big Three automobile manufac-turers. And, a wide range of 4150-type carburetors is offered for aftermarket use. Vacuum-secondary types are available in 390 through 850 cfm in stair-step sizes to fit almost any requirement. Double-pumper types with mechanical secondary actuation are available in 50 cfm increments from 600 through 850 cfm. A 390 cfm double-pumper is required for NASCAR racing.

A Pro-Series 4150 is available with 950 cfm airflow.

4150s have either single or double accelerator pumps. Any of three fuel bowls may be used, depending on requirements: side-hung, side-hung with external adjustment for level setting, or center-pivot with external adjustment for level setting and an inlet at each fuel bowl.

A 4150 with side-hung float bowls can be converted to center-pivot fuel bowls. Or, center-pivot or side-hung bowls can be adapted for off-road use by adding Part 34-3, Off-Road Fuel Bowls. Other items, such as bowl vents, 50cc high-capacity accelerator-pump kits and quick-change jet and secondary-spring kits, can also be installed.

A 4150 has two metering blocks. The secondary metering block has the same physical dimensions as the primary metering block. It is sandwiched between the secondary fuel bowl and the carburetor body—as is the primary metering block.

MODEL 4160

Why the 4160? It was engineered to be manufactured at lower cost than the 4150. In some instances, this lower cost, coupled with Holley's performance image, allowed car makers to put a Holley 4160 on high-performance engines. 4160s have been used by American Motors, Chevrolet, Chrysler and Ford as original-equipment. They are just as high-performance oriented as any 4150.

The 4160 has a *secondary metering plate* with built-in orifices screw-attached to the carburetor main body. This metering plate requires less drilling, no plugging or tapping, and no screw-in main jets or power valve.

The 4150 and the 4160 are absolutely alike in operation and constuction except that the 4160 does not have a secondary metering block. Because of this, the 4160 is 5/8-in. shorter than the 4150.

Some 4160s are factory-equipped with a thin steel plate sandwiched between the bowl gasket and the metering plate, see photo on page 51. This eliminates the possible performance deterioration that can happen if the combination metering plate and bowl gasket swells into the secondary metering-plate channels, reducing their capacity. On rebuilds you can add Metering Body Plate 1034-1993 against the large gasket and a paper gasket such as 108-27 between the steel plate and the secondary metering plate.

All 4160s have single accelerator pumps. Any of three fuel bowls may be used, depending on requirements: side-hung, side-hung with external adjustment for level setting, or center-pivot with external adjustment for level setting and an inlet at each fuel bowl.

A 4160 with side-hung float bowls can be converted to a 4150 with Kit 34-6. Kit 34-13 is used with the 0-3310 or almost any 4160 with center-pivot fuel bowls. This includes a secondary metering block with gaskets, a longer fuel-transfer tube and longer fuel-bowl screws. Other items, such as bowl vents, 50cc high-capacity accelerator-pump kits and quick-change jet and secondary-spring kits, can also be installed.

4160 carburetors have vacuum-

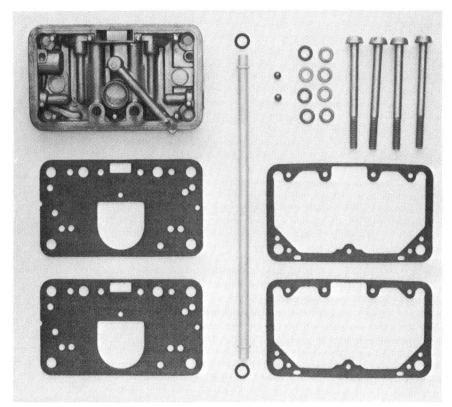

Kit 34-6 converts 4160s to 4150 secondary metering block configuration with replaceable main jets. Longer bowl screws, metering block, gaskets and a longer fuel-transfer tube are included. Lead balls close holes left when balance tube is removed. Note that a power valve cannot be installed in the metering block.

operated (diaphragm) secondaries. None have had mechanical linkage except the 0-4224 "center-squirter" 660 cfm for tunnel-ram manifolds (no longer offered).

For all-out racing, two of the shorter 4160s can fit onto a 2 X 4 manifold. Holley offers a Secondary Vacuum Balance Kit, Part 20-28 to adapt 4160s for 2 X 4 manifold applications.

If you need an aftermarket carburetor for your vehicle, you can buy a 4160 at a lower price than a 4150.

MODEL 4180

The 4180, a 4160 look-alike, is an emissions-oriented carburetor. It incorporates substantial changes to improve fuel metering, calibration and control. Most obvious are the *annular-discharge boosters*.

The primary metering block is substantially changed. To ensure idle quality, idle and main-well pas-

sages are drilled into the metering block (instead of formed cavities between metering block and main body gasket). Because of larger drilled wells, accelerator-pump passage routing had to be changed. And the primary main jets had to be angled into the metering block.

A brass orifice on the fuel-bowl side of the accelerator-pump passage allows pump vapors to bleed off into the bowl, keeping the pump filled. While this causes a minor loss of pump shot, it tends to eliminate pump "sag" after a hot soak. You can adjust the idle speed, but there are no idle-mixture-adjustment screws on the primary metering block. They are set at Holley and sealed under plugs.

Secondaries are vacuum-actuated and a metering plate is screwed onto the back of the main casting like a 4160. A thin steel plate (Metering Body Plate 1034-1993, photo on page

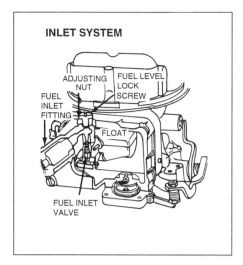

INLET SYSTEM

ADJUSTING NUT
FUEL LEVEL LOCK SCREW
FUEL INLET FITTING
FLOAT
FUEL INLET VALVE

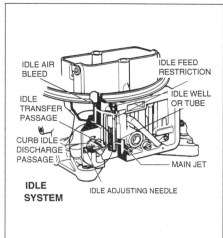

IDLE AIR BLEED
IDLE TRANSFER PASSAGE
CURB IDLE DISCHARGE PASSAGE
IDLE FEED RESTRICTION
IDLE WELL OR TUBE
MAIN JET
IDLE ADJUSTING NEEDLE
IDLE SYSTEM

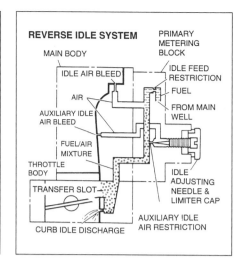

REVERSE IDLE SYSTEM
MAIN BODY
IDLE AIR BLEED
AIR
AUXILIARY IDLE AIR BLEED
FUEL/AIR MIXTURE
THROTTLE BODY
TRANSFER SLOT
CURB IDLE DISCHARGE
PRIMARY METERING BLOCK
IDLE FEED RESTRICTION
FUEL
FROM MAIN WELL
IDLE ADJUSTING NEEDLE & LIMITER CAP
AUXILIARY IDLE AIR RESTRICTION

51) is used between the metering plate and the bowl gasket so the gasket material can't swell into the metering-plate channels.

The 4180 was original equipment on the 1979-86 Ford 460 cid trucks, 1983-85 Ford/Mercury Mustang/Capri 302 cid, and 1984-87 Ford 351 cid trucks.

DESCRIPTION OF OPERATION

This description is applicable to all 2300, 3150, 3160, 4150, 4160, 4180, 4165, 4175 and 4500 (except for the intermediate system described in that section) and 2300 carburetors. No attempt has been made to explain why certain things happen in the functioning of the carburetor. These are fully covered in Chapter 2, *How The Carburetor Works*.

Fuel Inlet System—Fuel enters the primary fuel bowl through a screen or filter and passes through the inlet needle and seat assembly into the fuel bowl. Where there is only one fuel inlet for the carburetor, fuel reaches the secondary needle and seat through a transfer tube O-ring-sealed to each fuel bowl. If there are two fuel inlets, a fuel line must be plumbed to each. Holley's complete fuel lines make this easy. Although no longer used, a balance tube was sometimes included to connect primary and secondary bowls. This equalized fuel levels

when one inlet valve supplied more fuel than the other.

Idle—An idle system is provided for each barrel. The primary side is adjustable on all four-barrel units except the 4180, which has sealed idle adjustments. It is adjustable on the secondary side on competition carburetors. Holley's Street-Legal Carburetors have limited idle adjustment.

Except for adjustability, primary and secondary idle systems are essentially identical in construction and operation.

At idle, fuel flows from the bowl through the main jet and into the main well. From the main well, a passage connects to the idle well. Fuel flows through an idle-feed restriction into the idle well, up the idle well and is mixed with air from the air bleed as it flows down another vertical passage. At the bottom of this passage, idle fuel branches. One leg goes to the transfer slot above the throttle. The other goes past the adjustment needle (if used) to a discharge hole below the throttle plate. At curb idle, idle fuel is primarily supplied through the hole *below* the throttle.

Some 2300 and 4150/4160 carburetors have a reverse idle adjustment. Turn the screws in to richen the mixture and *back them out* to *lean* the mixture—just the opposite of what has always been done. This was done to improve idle quality with the lean idle mixtures need-

"Reverse" idle system is labeled. Adjustment is leaned by turning counter-clockwise, admitting more air, as shown in accompanying drawing.

ed to pass emission requirements.

The idle system was changed as shown in the drawing above. Note the adjustable second air bleed in this new idle circuit. The mixture screw varies inlet-air area through a second air bleed connected to the throttle bore just below the venturi. Idle improvement is so great that this system has been incorporated in other carburetors. It is used in 4160 and 2300 aftermarket carburetors for late Ford automobiles. It is also in most 4165, 4175 and in the emission/performance 4150 double-pumpers. Labels indicate the adjustment change.

Main Metering System—At cruising speeds, fuel flows from the fuel bowl, through the main jet into the bottom of the main well. Fuel moves up the well past air-bleed holes in the side of the well. These air-bleed holes are supplied with filtered air from the main air-bleed openings in the carburetor air inlet. The fuel/air mixture moves up the main well and across to the discharge nozzle in the booster venturi.

Power System—During high-speed operation or any situation when manifold vacuum is low (near atmospheric), the carburetor provides added fuel for power operation. A vacuum passage in the throttle body transmits manifold vacuum to the power-valve chamber in the main body. Vacuum applied to a diaphragm in the power valve holds the valve closed at idle and normal loads. When manifold vacuum drops, a spring inside the power valve opens the valve to admit extra fuel. Fuel flows through the power valve through the power-valve restriction and into the main well where it joins fuel flow in the main metering system to enrich the mixture.

As engine power demands are reduced, manifold vacuum increases. Power-valve-spring tension is overcome and the valve closes, stopping the extra fuel flow.

Some 4150s have two power valves, one for the primary and one for the secondary side.

WARNING: A power-valve diaphragm can be ruptured by an engine backfire. The result is an excessively rich mixture at all speeds. In 1992 Holley began incorporating power-valve blow-out protection in the throttle bodies of some 2300, 4150/4160/4180 and 4165/4175 carburetors.

You can add this backfire protection to older carburetors.

The CENTEK Power Valve Shield replaces one throttle-body baseplate screw with a special screw incorporating a check ball. The screw for the front power valve can be installed without any drilling or disassembly of the carburetor. The normal hole that supplied manifold vacuum to the power valve is

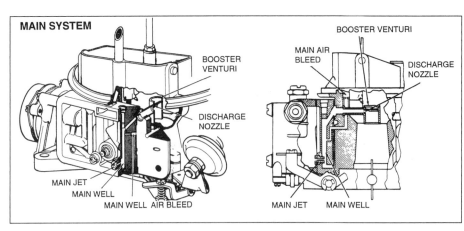

Annular booster shown with its two component parts. Top section containing the annular track and discharge holes is pressed into the larger ring section. Larger section contains the fuel discharge channel from the main well. Assembly is spun into place in the main body just like a conventional booster is installed.

Typical installation of annular boosters is seen in 0-9379, a 750 cfm competition carburetor with universal linkage, no choke, adjustable idle on secondaries and dual-inlet center-pivot float bowls.

Screw-in power valves with their respective gaskets. Window-type uses plain gasket without protrusions. Power valves with drilled holes use protrusions on the gasket i.d. to center the gasket. Drilled power valves were not used after the 1980s, but occasionally show up in new carburetors.

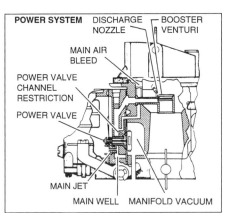

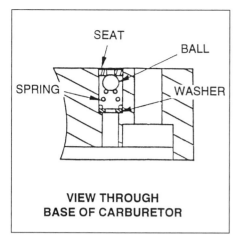

VIEW THROUGH
BASE OF CARBURETOR

Cross-section drawing shows construction of power-valve backfire protection factory-installed by Holley in throttle baseplate.

CENTEK's Power-Valve Shield is a throttle baseplate screw with a check-ball. It can be installed by simply replacing one baseplate screw and driving a plug into the hole that applied manifold vacuum to the power valve chamber.

closed off with a drive-in plug. CENTEK kits are available from performance warehouse distributors, carburetor shops and speed shops.

Accelerator Pump System—The pump is a diaphragm type in the bottom of the primary fuel bowl. There is also a pump in the bottom of the secondary fuel bowl on double-pumpers. The pump functions when the throttle linkage actuates the pump lever. Pressure forces the pump-inlet check valve onto its seat, preventing fuel from flowing back into the fuel bowl. Fuel from the pump is pushed through a diagonal passage in the metering block to the main body. Fuel under pressure raises a discharge check needle off its seat and fuel sprays into the venturi through the discharge nozzle.

As the throttle returns to a more-closed position, the pump linkage returns to its original position. The diaphragm spring forces the diaphragm down. The pump-inlet valve falls off its seat and the pump refills. As pressure is relieved from the pump passage, the discharge check needle reseats. Air flowing past the discharge nozzles will not pull fuel out of the pump passages.

Secondary System—Secondary systems are operated by a vacuum diaphragm or by a mechanical linkage to the primary throttles.

The secondary side has a fuel-inlet system, an idle system (usually non-adjustable) and a main metering system. Double-pumpers also have an accelerator-pump system. These systems operate exactly like those on the primary side in nearly all instances. Some secondary systems have a power valve, others do not.

Mechanical Secondary—There are various linkage types. Most use a direct link between the primary and secondary. A slot in the secondary lever provides progressive action. Through 1975 a slotted primary lever with a roller/lever transmitted opening motion through a rod to the secondary throttles. This method of secondary actuation has been retained on some competition carburetors, such as the 0 8156 and 0-8162. Mechanical secondary opening is controlled entirely by primary-throttle position. On some competition carburetors, the linkage is direct or 1:1. The primary and secondary throttles open equally at the same time.

When progressive action is used, the primary throttles open approximately 40° before the secondaries start to open. Both primary and secondary throttles reach wide-open at the same time.

Choke System—Both integral and divorced-choke mechanisms have been supplied on 4150/4160/4180 carburetors. Some have manual chokes.

Conversion kit 45-225 converts integral automatic chokes to manual operation. Some carburetors are factory-equipped with electric chokes. Kits 45-223 and 45-224 can be used to convert to this type of choke actuation. Kit 45-226 converts hot-air integral chokes to electric operation.

List 0-1850, 600 cfm—The lowest-cost four barrel 0-1850 is Holley's most-popular seller. This 4160 is designed for street/competition use on 1964 and earlier vehicles. It will not pass emission tests. Vacuum secondaries provide only the airflow which the engine requires, allowing successful use on 300 cid and larger engines. If you have a smaller engine, buy one of the 390 cfm (0-8007) or 450 cfm (0-4548) 4160 four-barrels.

The linkage works with GM and Ford engines except Ford automatic transmissions. Kit 20-48 allows use with Ford automatic transmissions with kickdown lever (except automatic overdrive). Throttle-lever extension, Part 20-7, adapts to Chrysler applications. A cable bracket, Part 20-95, works with GM's Turbo Hydra-matic 700R4 automatic transmission.

Cost-engineering at Holley has eliminated all non-essential parts to reduce cost without sacrificing performance. Some throttle linkage adapt-

0-9834 is a 600 cfm 4160 non-emission, street-performance carburetor cataloged for later-model—through 1979—non-EGR passenger-car and RV applications. Has electric choke, single swivel inlet for ease of plumbing, adjustable needle-and-seat assemblies in side-hung fuel bowls. Vacuum secondaries and emissions hookups combine with universal throttle linkage that accepts Ford's automatic-transmission kickdown linkage. A dashpot/solenoid bracket is included. It works well with a mild performance camshaft.

O-ring-sealed accelerator-pump transfer tubes were used on Holley four barrels for several years, starting in 1975. This is also used on the 4180. Gasket cutout clears tube. Any attempt to use the non-tube metering block with this cutout gasket will allow the accelerator-pump system to draw fuel from the bowl. A very-rich mixture will be supplied to the engine.

Second-design double-pumper mechanism has direct action. Slot in secondary lever takes up link travel during early primary opening and delays secondary opening. Secondary lever moves faster once it starts due to a smaller radius.

ing or other modifications may be needed which would not be required with a Holley bolt-on carburetor designed for a specific application.

Power-valve blow-out protection and all-metal parts were incorporated in 1992. Because the 0-1850 calibration is universal, some tuning of variables such as jets, power valve, secondary diaphragm spring, and so forth, may also be needed.

Fuel-line connection is simple. Install a 26-42 fitting and screw in a flared fitting. Or, use a swivel (banjo) fitting such as Holley's 26-25 with a 5/16-in. hose.

A hand choke with fast-idle adjustment can be replaced with an electric choke by using Kit 45-223. If your application requires an electric choke, or is for a Ford, consider buying a

Description of holes in main-body gasket surface facing metering block. These numbers match similar features in two metering blocks shown below.

1 — To discharge nozzle (accel.pump)
2 — To timed-spark port
3 — To curb-idle discharge
4 — To curb-idle transfer slot
5 — Used only with auxiliary idle air bleed
6 — To idle air bleed
7 — To main discharge nozzle
8 — To main air bleed
9 — Dowel locators for metering block
10 — Not used
11 — To bowl vent (pitot tube)
18 — Manifold vacuum chamber (for power valve operation)

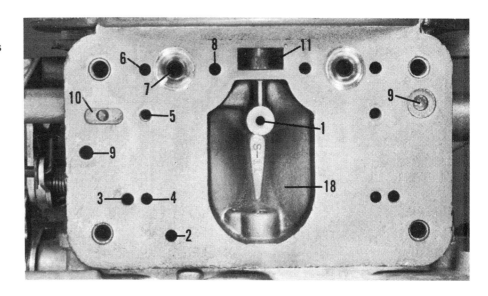

Description of holes and passages in metering block (two photos). Photo at right shows side which mates to main-body gasket. Lower photo is fuel bowl side of metering block.

1 — Accelerator-pump discharge passage
2 — Timed-spark passage (see 12)
3 — Curb idle discharge
4 — Idle-transfer fuel connects to main body and to curb idle adjust screw
6 — Idle bleed air enters from main body
7 — Main passage to discharge nozzle
8 — Main bleed air enters from main body
9 — Dowels to position bowl and gasket
11 — Bowl vent passage
12 — Timed spark tube boss
13 — Idle-mixture-adjustment needle
14 — Main jet
15 — Power valve threaded opening
16 — Power valve
17 — Power valve channel restriction (connects to main well 21)
18 — Manifold vacuum chamber (for power valve operation)
19 — Idle down well
20 — Idle well
21 — Main well
22 — Air bleed holes into main well
23 — Main air well
24 — Idle fuel from main well
25 — Idle feed restriction to idle well
26 — Fuel entry from accelerator pump in fuel bowl

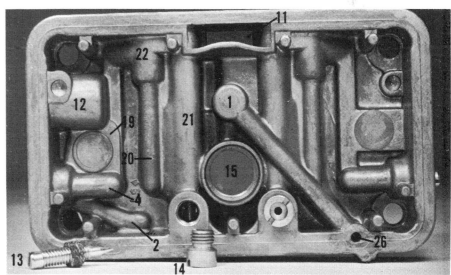

Holley's most-popular carburetor is the 0-1850. This 600 cfm street/competition, non-emission carburetor features vacuum secondaries, hand choke, single fuel inlet with side-hung fuel bowls and backfire-protected power valve.

34-1850 Quarter Mile Dial carburetor has driver-adjustable mixture control. A 0-1850 carburetor can be converted to this system with a 34-105 kit.

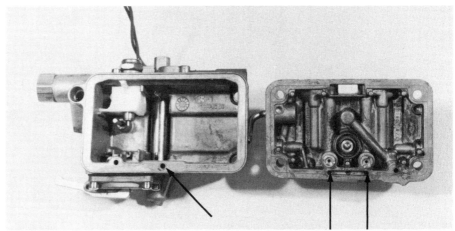

Electric choke kit 45-226 converts integral choke Holley carburetors to electric operation for use on engines with headers or without exhaust heat. Solid-state heating element is used.

Quarter Mile Dial float bowl with solenoid (float removed for clarity). Fuel is fed from bowl through solenoid jet via hole (arrow) to metering block. Metering-block slot distributes fuel through holes (arrows) to main wells.

0-9834 because it fits the application better. It includes an electric choke, Ford automatic-transmission kickdown linkage and a swivel (banjo) fuel-inlet fitting. It is calibrated somewhat leaner for use on 1974-82 cars and trucks and includes a dashpot/solenoid bracket.

Kit 34-105 converts your 0-1850 carburetor for use with the Holley Quarter Mile Dial System. This system allows you to change the fuel/air mixture from a dashboard-mounted control. The 34-1850 is a 0-1850 carburetor incorporating all of the specially designed components and circuitry of the Holley Quarter Mile Dial System.

List 0-3310, 750 cfm—This very popular four barrel is for 350 cid and larger engines.

Originally introduced in 1966 on Chevrolet's 425-hp Corvettes and the 375-hp Chevelles with 396 cid engines, the carburetor was also used as original equipment on 302 and 350 cid Z-28 Camaros.

Chevrolet versions of the carburetor have different numbers. Also, in 3310-1 form, it was a 4150. As the 3310-2, -3 and -4 it was changed to a 4160 to reduce cost. Earlier versions flowed 780 cfm.

Perhaps the reason for the 0-3310's popularity is due to its high-performance image. And, it installs easily

1992 and later 750 cfm Model 4160, List 3310 have all-metal parts, capscrews retaining fuel bowls, cast-in bowl vents, a new configuration on the front of the airhorn, universal throttle linkage and backfire-protected power valve. Vacuum secondaries and dual center-hung float bowls are standard features. This carburetor rates No. 2 in the Holley popularity contest.

on a wide variety of cars with good assurance of working well on almost any 300 to 455 cid engine. Its flexibility of application stems from diaphragm-operated secondaries which allow the carburetor to operate effectively over a wide range of airflow requirements.

This 750 cfm 4160 is designed for street/competition use. It will not pass emission tests. The throttle linkage works with GM and Ford engines except Ford automatic overdrive transmissions. Kit 20-49 converts the carburetor for Ford automatic transmissions with kickdown lever (except automatic overdrive). Throttle-lever extension, Part 20-7, adapts the 0-3310 to Chrysler applications. A cable bracket, Part 20-95, works with

GM's Turbo Hydra-matic 700R4 automatic transmission.

Holley eliminated all non-essential parts to reduce cost without sacrificing performance. In 1992 power-valve blow-out protection and all-metal parts were added. It is not a bolt-on replacement carburetor. Because the calibration is universal, some tuning of variables such as jets, power valve, secondary diaphragm spring, etc., may be needed.

Fuel-line connection is simple. Just use a Holley 34-160 fuel line to plumb fuel to the two inlet fittings.

A hand choke with fast-idle adjustment can be replaced with an electric choke by using Kit 45-223.

NOTE: If your application is for a Ford, buy 0-83310 instead because it

includes the Ford A/T linkage and has all of the other features of the 0-3310.

Or, if your Ford application can use a 600 cfm carburetor, buy the 0-9834 because it fits the application even better. It includes an electric choke, Ford A/T kickdown linkage, dashpot/solenoid bracket and a swivel (banjo) fuel-inlet fitting.

Kit 34-104 allows you to convert your 0-3310 carburetor for use with the Holley Quarter Mile Dial System. This system allows you to change the fuel/air mixture from a dashboard-mounted control. 34-3310 is a 0-3310 carburetor which incorporates all of the specially designed components and circuitry of the Holley Quarter Mile Dial System.

ADAPTERS & MANIFOLDS

It is always best to install a carburetor which fits the manifold correctly. 4150/4160/4180 and 4010 carburetors have a 5-3/16 X 5-5/8-inch bolt-hole pattern, sometimes referred to as a *square* flange or pattern. A single four-barrel intake manifold with the same pattern should be installed on the car. Otherwise, an adapter is required. Unfortunately, adapters are "thieves" that steal airflow capacity and may cause uneven fuel/air distribution. As an example, a 750 cfm 0-3310 flows only 640 cfm when adapted to a spread-bore manifold.

If your engine was originally equipped with a spread-bore carburetor (1-3/8 in. primaries and 2-1/4 in. secondaries), choose a Holley 4165, 4175 or 4011. These fit without an adapter. And, the fuel distribution is considerably better than can be expected when a square-bore carburetor is adapted to a spread-bore manifold.

If one of these is not offered for your car, consider changing to a manifold with the square-bore pattern. Adapting a carburetor is not an easy task because there are so many connections and linkages to consider.

Compare 4165 metering block (also 4175 primary) at bottom compared with 4150 block (top) shows housing for accelerator-pump discharge check ball as indicated by pen. 4165 block has two aluminum plugs on underside.

Small-primary/large-secondary throttle-bore concept is the best plan for staged carburetors, whether high-performance-oriented or not. Holley's 4165 series was the first spread-bore to combine high-performance capabilities and specification emission performance. Note two accelerator pumps and mechanical secondary linkage. These carburetors are bolt-on replacements for GM cars.

QUARTER MILE DIAL & MILE DIAL

Driver control of the fuel mixture is a reality. You can electronically adjust within a range of about five jet sizes at wide-open throttle.

Model 2300 and 4150/4160 carburetors with all the specially designed Quarter Mile Dial™ components and circuitry installed are referred to as *systems*. Fuel from the duty-cycle solenoid is directed to the main well. Mid-range and top-end performance is maximized. Kits are also available to convert popular Model 2300 and 4150/4160 carburetors to the Quarter Mile Dial system.

Mile Dial Carburetors™ and kits convert popular Model 2300 and 4150/4160 carburetors direct fuel from the duty-cycle solenoid to the power-valve circuit. This system allows adjustments for economy and other driving variables. Kits are also available to convert popular Model 2300 and 4150/4160 carburetors to the Mile Dial system.

MODEL 4165

Holley's spread-bore 4165 carburetors were introduced in 1971. These bolt-on replacements found instant favor among owners of cars equipped with Rochester Quadra-Jet (Q-Jet) and Carter Thermo-Quad carburetors.

A true "bolt-on" accessory or part is rare in the automotive business, as you know if you have done much wrenching. Holley engineers created a carburetor which bolts to the manifold and retains all stock equipment, including air cleaner, linkage and all emission-control connections.

Initial development was done on Chevrolets. The 650 cfm unit was calibrated on a 350 cid engine and the 800 cfm unit developed from tests on both 402 and 454 cid engines. Under actual driving conditions, the Model 4165 outperformed the original carburetor in traffic accelerations, hot starting, hot-idle conditions, vapor-lock prevention and general driving. In basic wide-open-throttle acceleration tests on a 350 cid automatic transmission Impala, the Model 4165 was 0.85 seconds and 5.4 mph faster over a 1/4-mile drag strip. Dynamometer testing showed 14-hp performance improvements on both small and big-block Chevys. Emission levels were similar to the original equipment.

Holley supplies Street-Legal Direct-Replacement 4165s for Buick, Chevrolet, Chrysler, Dodge, GMC, Oldsmobile, Plymouth and Pontiac. In some cases they replace the Rochester-Q-jet and in others the Carter Thermo-Quad..

A 650 cfm 4175 model is similar to the 4165 except it has vacuum-operated secondaries, side-hung fuel bowls and a single accelerator pump. It is discussed later.

Diaphragm-operated secondary throttles make this Model 4175 ideal for spread-bore applications where heavy loads occur at low speeds, as with trucks and camper-type vehicles. This is the 0-9895, a 650 cfm for Chevrolet 1975-78 L-48 and L-82 350 cid engines.

4165 Features

At first glance the 4165 carburetor looks like any other Holley four-barrel of the 4150/4160 double-pumper class. It's not! Very little is the same on the 4165, except for jets, power valves, bowl vent "whistles," needle and seat assemblies and floats. Gaskets, metering blocks, accelerator pumps, pump-discharge nozzles (shooters) and fuel bowls are all different. They don't interchange with other Holley parts.

Let's look at the 4165 to see how it is made and why it's different. First of all, the bolt holes are on a 4-1/4-inch X 5-5/8-inch spacing to fit spread-bore manifolds. Next, the carburetor has the appropriate connections for emission equipment and the correct linkage to fit the specific automobile using that list number. Some 4165s for street/competition/off-road use may not allow hooking up all emission devices.

Primary & Secondary Venturis— One important difference between the 4165 and original-equipment carburetors is the use of fixed-size secondary venturis and removable main jets. Primary 1-5/32-in. diameter venturis provide good response during part-throttle operation and allow a wide range of vehicle speeds and load conditions without operating the secondaries. Keeping "out of the secondaries," achieves greater fuel economy.

Primary venturis are smaller than the secondaries to allow meeting emissions requirements. A small (20 cc per 10 strokes) accelerator pump on the primaries matches primary-venturi airflow requirements.

Secondary venturis are either 1-3/8-in. or 1-23/32-in. diameter, depending on whether the carburetor is a 650 or an 800 cfm unit.

Secondaries are mechanically operated and use a high-capacity (50 cc per 10 strokes) accelerator pump.

Accelerator-Pump Circuits— Because these are different on the 4165, the bowl gaskets, metering blocks, fuel bowls and discharge nozzles are unlike other Holley carburetors. Pump circuits are the same for primary and secondary. These were the first to use Holley's rubber type inlet/check valves. A steel-ball inlet/check was used on the secondary side of some early 4165s.

The rubber "umbrella" inlet valves came from an extensive testing program to get good hot-fuel-handling characteristics. When the carburetor and fuel are hot, an off-idle bog will occur if the pump has "dried out." Fuel in the pump cavity boils away (it's close to the engine and subject to a lot of heat) during a hot soak. Several features were incorporated to enhance hot-start driveaway capabilities. Both pumps (primary and secondary) are self-filling so they always fill immediately after the pumps have been used.

Model 4175 0-80073 is a 650 cfm direct replacement for Rochester Quadra-Jet on 1982-83 Chevrolet Z-28 and Pontiac Trans-Am with 305-cid engines. Vacuum secondaries. Full emissions provisions and direct computer hookup. Feedback solenoid is at front of primary bowl. Also see photos at right.

It's very important for the primary pump to provide an instant fuel shot at the slightest throttle movement.

The inlet/check valve seals instantly as the pump operates from a very small throttle movement. The discharge check valve is a steel ball near the bottom of the pump passage. This check, in conjunction with an anti-pullover discharge nozzle (to prevent fuel in the nozzle passage from being pulled into the airstream), tends to keep the passage to the "shooter" (nozzle) filled with liquid fuel. As the throttle is moves back toward a closed position, the ball discharge check seats instantly to create a partial vacuum in the pump cavity, further ensuring quick pump filling. The lightweight ball check also allows vapors to escape from the pump cavity during hot-soaks.

Discharge nozzles are targeted so the pump shot "breaks" against the lower edge of the booster venturis. This ensures at least partially vaporized fuel entering the engine.

All 4165s are double-pumpers.

Power Valves—Power valves are the same as used in the 4150/4160 carburetors. Backfire protection for the power valve was added in 1992.

Reverse Idle—Some 4165s have "reverse" idle adjustment as described and illustrated on page 53. The adjust-ment screws in to richen the mixture and backs out to lean it—opposite of what has always been done in the past. Labels identify this changed idle system.

Standard & High-Performance Types—There are two versions of 4165 carburetors: Street-Legal Direct-Replacement and Street Legal, Performance Replacement. 4165 carburetors are "high-performance" in so many respects that it's strange to apply other names. These carburetors automatically give a performance increase while holding emissions at legal levels. And mechanical secondaries and double accelerator pumps are in the same package.

Standard units have a single fuel inlet on the front float bowl. Side-hung float bowls are connected by a fuel-transfer tube that is O-ring-sealed where it enters each fuel bowl. Float resetting requires removing the fuel bowls and bending a tab on each float.

A brass baffle on the metering block in the primary bowl reduces fuel slosh through the vent into the carburetor on acceleration. Fuel bowls on the standard carburetors allow using stock non-high-performance air cleaners. If the air cleaner must be rotated slightly for installation, the air-cleaner-base locating tabs may have to be modified.

Automatic chokes are provided on standard 4165s. Buick, Chevrolet and Pontiac units through 1972 use the stock divorced-choke actuating mechanism; 1973 and later Pontiac and Oldsmobile units use an integral choke with an accessory stove with connecting tubing that is installed on the intake manifold heat riser on some models. Chrysler units also work with the stock divorced-choke equipment. Kit 45-518 adapts divorced-choke 4165/75s to aftermarket manifolds.

High-performance "Street/competition/off-road non-emission" 4165s are exactly like the "standard" units except the fuel bowls are the center-pivot-float type with externally-adjustable sight plugs on the side of the bowls for setting fuel level. Two other differences are in the use of "whistle" type bowl vents in both primary and secondary bowls and a manual choke linkage.

If you buy a high-performance version you may have to replace the stock air cleaner with a high-performance unit to clear the center-pivot "race" bowls. An accessory fuel line is needed to plumb fuel to both bowls and you must hook up a manual choke to complete the installation. Linkage on these carburetors fits the cars for which each carburetor is designed.

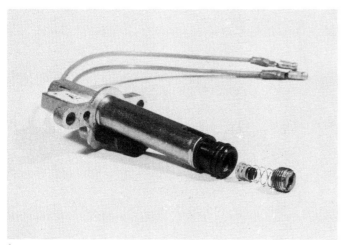

Solenoid from 4175 with poppet valve, spring and main jet removed. Main jet and poppet are not serviced separately. Jet size is factory-selected so carburetor will meet emission requirements.

Primary metering block of closed-loop Model 4175 (electronic) has slot with three holes under the main jets. This area is fed supplementary fuel under electronic control via the solenoid valve in the primary fuel bowl.

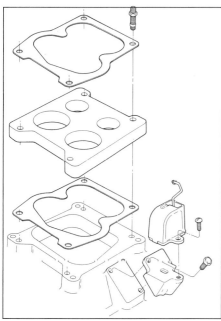

Kit 45-518 adapts 4165/75 divorced-choke carburetors to aftermarket manifolds.

650 cfm versions are recommended for all normal-replacement applications from 327 through 402 cid engines. The same carburetors also work very well on low-performance "smog" version 427 and 454 cid engines. 800 cfm units are recommended for standard replacement on high-performance 396, 402, 427 and 454 engines. They can also be used on racing-type small-block engines with a real need for this much airflow capacity.

MODEL 4175

In 1974 Holley introduced the Model 4175. It was designed specifically for pick-up trucks, campers, race-car tow vehicles and other applications on 340 cid and larger engines where low-speed "lugging" makes mechanical secondaries undesirable. Flow capacity is 650 cfm. Divorced and integral chokes are supplied so the carburetor works with the stock equipment on Chevrolet, Oldsmobile, Pontiac and Chrysler applications. It is exactly the same as the Model 4165 except it has diaphragm-operated secondaries. There is no accelerator pump on the secondary side. A metering plate is used on the secondary side.

Computer-Controlled Model 4175 for GM Cars/Trucks

List 0-80073 utilizes the same duty-cycle solenoid and related hardware as used in Holley's Quarter Mile Dial and Mile Dial Systems. But instead of the driver controlling the duty-cycle manually, it connects directly to the GM electronic control system. This carburetor is a direct plug-compatible replacement for Quadra-Jet-equipped 305 cid Chevrolet/GMC 1983 trucks, 1982-83 Pontiac TransAm and Z-28 Camaro. It has vacuum secondaries, flows 650 cfm and has full emissions hook-ups.

This 4175 is totally compatible with GM systems. Metering specifications were calibrated to suit. A few physical characteristics and mechanical hook-ups were modified to work with various GM cars. And that good response and "feel" for which Holley is so well known has been retained.

Three members of the Model 4500 family: 0-7320, 0-8082 and 0-8896 (left to right). Airflow capacity is 1150, 1050 and 1050 cfm respectively. Two others, 0-9375 and 0-9377 offer 1050 and 1150 cfm airflow respectively and feature annular discharge boosters. A 750 cfm unit, 0-80186 (shown below), also has annular-discharge boosters.

MODEL 4500

The first Holley 4500 was designed for Ford in 1969 as a single four barrel on NASCAR Stock Car engines. This expanded four barrel mated a sand-cast carburetor body to the basic Holley metering blocks and fuel bowls. Base stud spacings allowed 2-inch-diameter throttle bores and 1150 cfm airflow capacity. An accelerator pump on each fuel bowl made these some of the first "double-pumpers."

Design Features

Several versions are separately described in this section, but some features are common throughout. All have 2.00-inch throttle bores and 5.380-inch-square stud pattern. Dual-inlet center-pivot-float fuel bowls are tapped on both sides to allow fuel-line connections to either side. Plastic bowl vents are provided in all fuel bowls. Reduced-section 1/2-inch throttle shafts are used for improved airflow. The "blind end" of each shaft is sealed with an expansion plug. The lever end is sealed with an O-ring. Linkage between primary and secondary throttle shafts is enclosed inside the carburetor body to protect it from dirt.

Each of the rugged 0.150-inch-thick throttle-shaft levers has a plastic

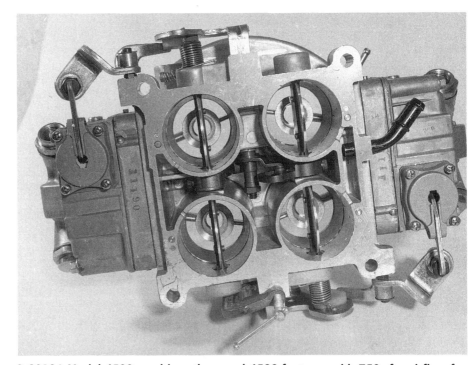

0-80186 Model 4500 combines the usual 4500 features with 750 cfm airflow for smaller engines. Booster "tails" reduce effective venturi diameter to reduce the airflow. Throttles held open with clamp for photo.

cam to actuate an accelerator pump. Accelerator pumps give 30cc per 10 strokes on the 6214 I.R. type; 50 cc per 10 strokes on all of the other 4500s. Both accelerator pumps fill through rubber-type inlet valves.

The 4500 carburetor should only be used on a manifold designed with 5.380-inch-square stud spacing pat-tern. Any adapter to fit it onto a smaller manifold seriously reduces airflow capability.

Air cleaners for the 7-1/4-inch mounting pad of the 4500 are available as Part 120-4500 (paper element) and 120-4501 (foam element). K & N Engineering also makes air cleaners to fit the Model 4500.

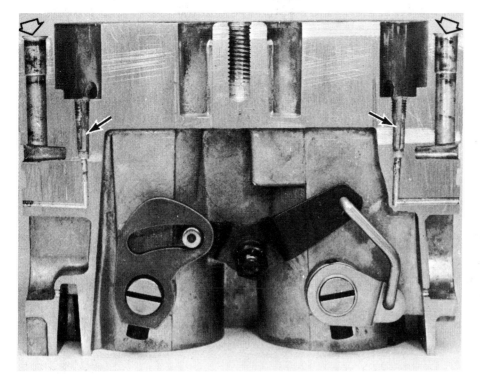

End view of 0-7320 shows 50cc per 10 strokes accelerator pump used on both bowls of all current production Model 4500s. Center-hung race bowl with externally adjustable needle-seat has dual inlets for plumbing fuel lines from either side. These fuel bowls are on all 4500s.

Saw a 4500 in half and you'll see a rugged linkage that's fully enclosed once carburetor is on manifold. This is the 1:1 linkage on a 6214. Black arrows indicate accelerator-pump discharge passages. Outline arrows point to vent-tube bosses for bowl venting through main body.

Many Pro-Stock engines are based on the venerable Chevrolet big-block, no matter what they may be called. Most of the engine builders use fabricated manifolds, as seen in the left photo. Right photo shows a Ford with a cast tunnel-ram manifold.

Three-inch velocity stacks are offered. These chrome stacks are typically used on the 6214 and 0-9377 to contain fuel standoff—and for their good looks. A slot must be cut into in the base of each stack to allow fitting over the booster-nozzle assembly and to provide a passage for fuel from the pump shooter into the venturi. Stacks are held by a long bolt with a metal safety tab. Velocity stacks can be installed after removing any choke.

Because these are essentially "stretched" versions of the 4150/4160 carburetor, look the section starting on page 53 for a description of operation. Only the intermediate-idle system is different. It is fully detailed in the idle system portion of Chapter 2, *How The Carburetor Works*, page 19. All 4500s are double pumpers. Some have no power valves, others have one and two have two power valves.

MODEL 4500 COMPARISON CHART

	Application	cfm Flow	Venturi (in.)	MJ	P.V.	Idle	Linkage[2]	Inter. Idle System	Choke	Accel. Pump	Booster[3]	Cam
4575[1]	General	1050	1-11/16	84	6.5	Prim only	Prog.	No	Yes	50cc	Short	Yellow
6214[1]	I.R. (Obsolete)	1150	1-13/16	95	NA4	4	1:1	Yes	No	30cc	Long	White
6464[1]	Pro-Stock Plenum ram 2x4 (Obsolete)	1050	1-11/16	88	NA4	4	1:1	Yes	No	50cc	Short	Yellow
0-7320	Plenum ram 2x4 or single 4-bbl.	1150	1-13/16	95	Plugs[5]	4	Prog.	No	No	50cc	Short	Yellow
0-8082	Same as 0-7320	1050	1-11/16	84	5.6[5]	4	Prog.	No	No	50cc	Short	Yellow
0-8896	1X4 with auto. trans. 2X4 mini-plenum	11050	1-11/16	88	Plugs	4	Soft-Prog.	Yes	No	50cc	Short	Yellow
0-9375	2X4 mini-plenum	1050	1-11/16	86	None	4	Soft-Prog.	Yes	No	50cc	Annular discharge	Yellow
0-9377	2X4 mini-plenum	1050	1-11/16	94	None[4]	4	Soft-Prog.	Yes	No	50cc	Annular discharge	Yellow
0-80186	Street, 1X4 or 2X4	750	1-11/16	65	6.5[4]	4	Soft Prog.	No	No	50cc	Annular discharge	Yellow

Notes:
[1]Obsolete part number, no longer made.
[2]Replacement linkage cams are available for 1:1, soft progressive and progressive operation, See Holley Performance Parts Catalog.
[3]Kit 34-9 Conversion Kit has annular-discharge booster venturis, retainer sleeves and all parts necessary to update any model 4500.
[4]Kit 34-8 includes metering blocks to convert 0-6214, 0-6464 and 0-9377 to use power valves in primary and secondary.
[5]PV restriction passages are machined. Supplied with PV plugs. If larger than 100 main jets required, remove plugs and install a power valve. Then reduce main-jet size.

Annular-booster Kit 34-9 converts Model 4500s to annular-booster configuration. Brass inserts are inserted from metering-block-gasket surfaces to lock boosters in position.

ANNULAR DISCHARGE BOOSTERS

In 1980 Holley introduced a concept never before used in performance carburetors: annular-discharge boosters.

Annular-discharge boosters have higher air velocity close to the venturi surface where the discharge holes are located. This increases the metering signal and gets the main system flowing at lower airflows. The higher signal, together with equally spaced radial location of the discharge holes improves atomization and cylinder-to-cylinder distribution. The higher signal reduces main-jet requirement by 4 to 5 sizes. With annular-discharge boosters there is less dependence on accelerator pumps to cover up hesitation or bogs.

Annular-discharge boosters are used in the Model 4180 and in 4150 competition carburetors 0-9379 (750 cfm), 0-9381 (830 cfm) and 0-9380 (850 cfm). They are also used in Models 2010, 4010, 4011 and 4500.

Underside shows rugged enclosed linkage connecting primary and secondary throttles. Accelerator pumps are actuated by levers operating on cams at the back of the throttle lever and on the secondary shaft (arrows).

4500 with Holley air cleaner installed. Both paper and foam element filters are available.

Accelerator pump uses umbrella-type inlet valve (arrow) described on pages 33-35. This ensures a quick pump shot when the throttle is actuated.

Holes feed fuel from bowl for intermediate-idle system (arrows). Whistle-type vent is standard on 4500s.

Slot and discharge nozzle for intermediate-idle system (arrows), see pages 32-33 for description.

Velocity stacks, Part 17-16, add 3-in. rams to Model 4500. Center keeper holds stacks in place and locks securing capscrew.

Three Holley 2010 carburetors, left to right, are a 560 cfm (no choke), 500 cfm and a 350 cfm. All have universal throttle linkage with Ford A/T kickdown, standard Holley two-barrel flange, annular discharge nozzles and a 5-inch airhorn. An accessory electric choke is available.

Straight, right-angle and 45-degree fuel-inlet hose fittings are available for the 2010. Or, steel-line fittings can be screwed into the fuel-inlet fitting.

MODEL 2010

In 1992 Holley introduced a new two-barrel carburetor designed specifically for performance. These carburetors are dramatically simpler and lighter than the Model 2300 two-barrels. Venturis of 1-3/16 in. and 1-9/16 in. with 1-11/16 in. throttle bores provide 350 and 500 cfm airflows. The 560 cfm 0-82012 couples its 1-9/16-in. venturis with 1-3/4-in. throttle bores to get an additional 60 cfm.

These carburetors have a custom polished finish.

The 2010 carburetors are universal in application so they fit or can be adapted to almost any two-barrel manifold. Because the fuel bowl is an integral part of the throttle body, only one gasket is needed. Six Torx screws retain the airhorn/air-cleaner mount/fuel-bowl cover, simplifying main-jet changes. The power valve is accessible from the bottom of the carburetor

Adjustment and repair of the Model 2010, 4010 & 4011 carburetors is described in the Adjustment & Repair Chapter starting on page 145.

through a small cover attached with four Torx screws.

Annular-discharge boosters are used in all 2010s.

A number of components used in the 2010 are exactly the same as those also used in the 2300 and 4150/60/65/75/80 series carburetors: adjustable inlet needles and seats, power valves, main jets and lever-operated 30cc and 50cc pumps with umbrella-style inlet valves. Fuel bowls are equipped with center-pivot floats, sight plug and an externally adjustable needle-and-seat assembly.

Three 1/4-in. hose connections in the carburetor base are for timed and full manifold vacuum, spark advance and for an EGR valve. The timed port can also be used to purge charcoal canisters. There is also a 3/8-in. manifold-vacuum hose connection for PVC.

A manual choke is provided on the 350 and 500 cfm carburetors.

The base flange is the same as used on Holley's 2300: An SAE 1-1/2-in. spread-bore pattern (3-1/2-in X 5-1/2-in. hole spacing. This fits Ford two-barrel 289-390 cid engines.

The airhorn/float-bowl cover is the standard 5.0-in. that works with most performance air cleaners.

The universal throttle linkage has a Ford automatic transmission kickdown lever. It does not work with Ford automatic overdrive transmissions. Chrysler applications require extension lever 20-7.

The 2010 is literally the front half of a 4010/4011 carburetor, so its operation is just like the description for the Model 4010 and 4011 that follows.

MODEL 4010 & 4011

With all new automobiles now fuel-injected, you may ask why Holley introduced a complete new line of carburetors in the late 80s .

While new cars are typically powered by 4- and 6-cylinder engines and fuel injection, the V8 still rules the performance world. There's almost an unlimited supply of parts and equipment available and enthusiasts crave the performance that's only available with V8-equipped vehicles.

While Holley's 4150/4160 performance carburetors evolved from original-equipment carburetors, the 4010 and 4011 are truly new high-performance pieces.

There are two models, the 4010 and 4011. The 4010 is built on the typical Holley square flange; the 4011 on the Rochester Quadrajet spread-bore flange.

Bottom of 2010 showing throttle plates, accelerator pump, power-valve cover and vacuum hookups.

Top view of 560 cfm Model 2010 (0-82012) shows 1-9/16-in. venturis, no choke. Throttle bores are 1-3/4 in. This makes a great competition carburetor for two-barrel circle-track classes.

Universal throttle linkage on 2010 works with Ford A/T kickdown. A Chrysler throttle-lever extension is available.

From a design and function point of view, the two are essentially the same. But because the two flange arrangements are so different there is quite a physical difference. Nonetheless, think of them as one carburetor with some obvious flange differences.

Each model comes in two different configurations; a mechanically operated secondary and a diaphragm-operated secondary. Those with mechanically operated secondaries are double-pumpers. Each configuration comes in two different airflow ratings: 600 and 750 cfm for the 4010; 650 and 800 cfm for the 4011. Electric chokes are offered on each model. There are marine versions of each model.

General Description

Extremely simple design is used throughout: The main body and throttle body are combined into a single polished-aluminum casting containing fuel bowls, venturis, throttle bores and accelerator-pump cavity.

The airhorn, also polished aluminum, acts as a cover for the bowls, provides a choke housing and a 5-inch-diameter mounting surface for the air cleaner. Fuel inlets and float mounting are also in the airhorn.

Booster venturis are part of a cluster containing idle and high-speed bleeds, well tubes, idle-feed restrictions and accelerator-pump shooters. All clusters have an insert that provides annular discharge of the main-system fuel.

Most part numbers come with a manual choke, but conversion to an electric choke is easily done.

Although these are universal carburetors intended for the performance aftermarket, all part numbers provide hookups for emission equipment such as EGR, PCV and canister purge.

Advantages

Aluminum makes a very light carburetor. The average weight is about 6 pounds—about half that of a Model 4160 or 4165.

Because bolt-on versions are available for square-flange and spreadbore manifold configurations, there's no need for airflow-robbing adapters.

Fuel levels are externally adjustable, a great convenience for the tuner.

Primary and secondary bowls have their own fuel inlets. Polished-aluminum castings and black hardware create a very high-tech appearance.

Annular-discharge boost venturis provide a strong metering signal at lower airflows and prepare the mixture for combustion. Cylinder-to-cylinder distribution is also improved.

Main-jet changes are easily made with the removal of Torx screws and an airhorn stud. There's only one non-stick gasket. Just lift the airhorn and the fuel bowls, main jets and booster assemblies are exposed.

Main jets, power valves, accelerator-pump parts, diaphragm springs and inlet valves are all standard Holley parts that are easily obtainable.

These carburetors are easily modified. Remove screws holding the boosters to expose the main venturis, which can then be easily bored and polished.

Holley bolt-on kits and spare parts such as electric chokes, pump cams, diaphragm springs, power valves and main jets can be used on the 4010 and 4011 carburetors.

Description of Operation

The following descriptions are applicable to all 4010 and 4011 carburetors. And, with the exception of the secondary system, these descriptions also apply to the Model 2010 two-barrel. The basic functions of these carburetors' systems are fully covered in Chapter 2, *How The Carburetor Works*.

Fuel Inlet—Fuel under pressure enters through fuel fittings in the airhorn. Both primary and secondary

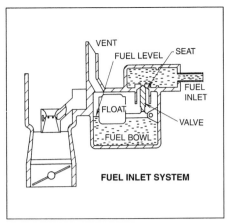

Fuel enters at fuel-inlet fitting. Flow into fuel bowl is controlled by inlet valve. Flow stops when float force overcomes pressure applied to needle by incoming fuel. As fuel level drops, needle opens to admit additional fuel. System maintains fuel level at a specific height regardless of flow. Float's physical position when valve is closed determines fuel level.

Fuel inlets on top of air-horn/float-bowl cover require angled fuel lines to clear air cleaner.

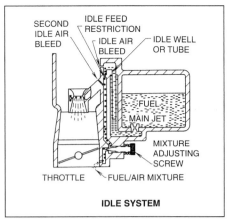

Idle-system flow is induced by manifold vacuum and controlled by idle-mixture-screw position. Fuel is drawn up idle tube where it meets air from first idle-air bleed. This mixture is then metered by the idle-feed restriction. More air is added by a second idle-air bleed. When throttle plate is opened farther, it exposes idle-transfer holes (not shown) increasing flow through idle system.

have their own individual fuel inlets. Inlet valves are the typical Holley externally adjustable type needle-and-seat assemblies.

The hollow plastic floats are hinged to the airhorn.

A combination airhorn float-bowl cover tops the fuel bowls which are an integral part of the main body. There are no gaskets between the bowls and main body and thus no potential leaks.

Idle System—An idle system is provided for each barrel. Most late-model original-equipment carburetors do not have an adjustable idle mixture. But idle adjustment is provided on the primary side because the 4010 and 4011 are universal carburetors designed to be used on different engine sizes and variations. Secondary idle mixture is not adjustable.

Idle speed is controlled by an adjustable idle stop that varies the throttle opening and, hence, the amount of air allowed to pass around the throttle plates.

Idle bleeds and restrictions are in the boost-venturi clusters (nozzle bars).

Idle fuel flows from the bowl through the main jets into the main well. Fuel is drawn up an idle tube in the cluster and is metered by a restriction at the top. Here the fuel meets air from the idle air bleed. This creates an emulsified fuel/air mixture that flows down a vertical passage. At the bottom of this passage, idle fuel branches. One branch goes to the transfer slot above the throttle, the other branch passes the adjustment needle on the way to a discharge port in the throttle bore.

At curb idle, fuel is primarily supplied through the hole below the throttle. As the primary throttle plate is opened, additional idle-transfer holes in the throttle bore are exposed so additional fuel and air are discharged. Proportionally more air is bypassed around the throttle plates so the mixture gradually leans and there is a smooth transition to part-throttle mixture.

Main Metering System—At cruising speeds, fuel flows from the fuel bowl through the main jets in the floor of the fuel bowl and into the main well.

Fuel moves up the well past air-bleed holes in the side of the emulsion tube. These air-bleed holes are supplied with filtered air from the main air bleed openings in the booster-venturi cluster. See accompanying drawing.

Fuel/air mixture moves up the main well and across to the annular-discharge nozzle in the booster venturi. An antisiphon bleed in parallel with the main air bleed prevents fuel flow from continuing once the booster signal is removed (as the throttle is closed to idle).

Each main venturi is cast integrally with the aluminum main body/throttle-body combination.

Power System—During high-speed operation or any driving condition when manifold vacuum is low (near atmospheric), the carburetor uses a power valve to add fuel for power operation. A vacuum passage in the main casting transmits manifold vacuum to the power-valve chamber.

Vacuum applied to a diaphragm in the power valve holds the valve closed at idle and normal loads. When manifold vacuum drops, a spring

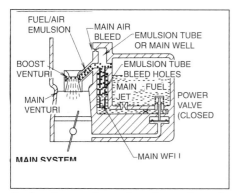

Fuel enters main system through main jets. The fuel-air combination, or emulsion continues to boost venturis where it joins incoming air. Low pressure in boost venturis creates a pressure drop—signal—necessary to induce flow. Incoming air flow is controlled by throttle opening.

Typical streetrod installation of Model 4010 on small-block Chevrolet by Bill Herbert of Tucson, Arizona. Dressup items include Holley valve covers and 9-inch air cleaner.

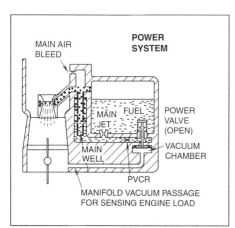

When load on engine increases, manifold vacuum drops. At a predetermined manifold vacuum, power-valve spring opens valve as shown. Fuel enters main system and is metered by power-valve-channel restriction. This fuel is supplemental to that entering through main jets, enriching the fuel/air mixture. Power valve is a gate or switch that permits flow. Power-valve-channel restrictions determine amount of fuel added when power valve opens.

inside the power valve opens the valve to admit extra fuel. Fuel flows through the power-valve restriction and into the main well where it joins fuel flow in the main metering system to enrich the mixture.

As engine power demands are reduced, increased manifold vacuum overcomes power-valve-spring tension. The valve closes, stopping the extra fuel flow.

Except for the 600 cfm versions of the Model 4010, both the primary and secondary sides of the 4010 and 4011 have power valves.

Power valves are the traditional power valves that have been used in Holley's other performance carburetors for many years.

Accelerator Pump System—The accelerator pump is a diaphragm type in the bottom of the primary fuel bowl. On double-pumpers there is a second pump in the bottom of the secondary fuel bowl. The pump functions when the throttle linkage actuates the pump lever. Pressure forces the rubber-umbrella-type pump-inlet valve onto its seat, preventing fuel from flowing back into the fuel bowl. Fuel from the pump is pushed past the open discharge check valve and into the cluster assembly. The pump-discharge restrictions, or shooters, are in the cluster assembly and are directed down into the incoming air stream.

Model 4010 has a 30cc pump on the primary side and no pump on the

secondary side of the vacuum-secondary version. The double-pumper version has a 30cc pump on both the primary and the secondary.

The model 4011 has a 30cc pump on the primary side of both versions. Because of its large airflow capacity, the secondary side of the double pumper has the larger 50cc pump.

Secondary System—This is operated by a vacuum diaphragm or by a progressive mechanical linkage to the primary throttles. The secondary side has a fuel-inlet system, a non-adjustable idle system and a main metering system. Double-pumpers also have an accelerator-pump system. The secondary system may also have a power valve.

Choke System—A manual choke is standard on most 2010, 4010 and 4011 carburetors. There is no choke on the Model 2010 560 cfm carburetor. At least one 4010 and one 4011 have electric chokes.

Kit 45-223 converts the 4010 to electric automatic choke. The 4011 uses the 45-459 for vacuum-secondary versions and 45-450 for the mechanical-secondary types. The 2010 can be electric-choke equipped with 45-459.

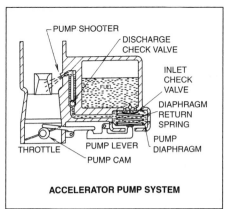

ACCELERATOR PUMP SYSTEM

PUMP SHOOTER
DISCHARGE CHECK VALVE
INLET CHECK VALVE
FUEL
DIAPHRAGM RETURN SPRING
THROTTLE
PUMP LEVER
PUMP DIAPHRAGM
PUMP CAM

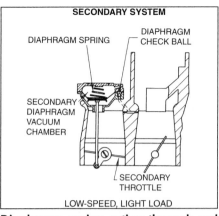

SECONDARY SYSTEM

DIAPHRAGM SPRING
DIAPHRAGM CHECK BALL
SECONDARY DIAPHRAGM VACUUM CHAMBER
SECONDARY THROTTLE

LOW-SPEED, LIGHT LOAD

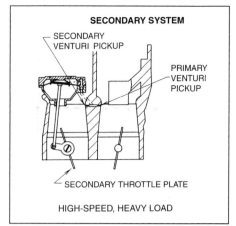

SECONDARY SYSTEM

SECONDARY VENTURI PICKUP
PRIMARY VENTURI PICKUP
SECONDARY THROTTLE PLATE

HIGH-SPEED, HEAVY LOAD

When throttle closes, spring forces accelerator-pump diaphragm down, drawing fuel is drawn into chamber through inlet check valve. Outlet check is closed during this process so air is not drawn in. Accelerator-pump system is now charged. When throttle is opened, fuel pressure closes inlet check and opens outlet (discharge) check valve. Fuel is forced into airstream through pump-discharge restrictions (shooters).

Diaphragm spring acting through rod keeps secondary throttles closed so all airflow is through primaries.

As mass air flow increases through primary side, vacuum increases in primary venturis. When vacuum becomes stong enough, it overcomes the diaphragm spring, allowing secondary throttles to begin opening. Air moving around ball check allows secondaries to open smoothly. When primaries are suddenly closed, ball is unseated, allowing secondary throttles to close quickly. Secondary venturis may bleed off some signal to help control opening rate.

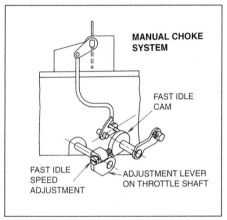

MANUAL CHOKE SYSTEM

FAST IDLE CAM
FAST IDLE SPEED ADJUSTMENT
ADJUSTMENT LEVER ON THROTTLE SHAFT

With a manual choke, choke plate is opened or closed through a linkage. Fast-idle cam allows for greater throttle opening when choke is fully or partially closed. Adjustment allows setting fast-idle rpm.

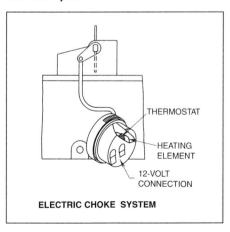

THERMOSTAT
HEATING ELEMENT
12-VOLT CONNECTION

ELECTRIC CHOKE SYSTEM

The 4010 and 4011 consist of two basic castings: fuel-bowl and throttle body combination and an airhorn/float-bowl cover. Also shown are primary and secondary clusters (nozzle bars) containing booster venturis and accelerator-pump shooters.

Force to open or close automatic choke comes from bi-metal spring that moves with a temperature change. Movement results because the two bonded-together metals have different expansion rates. Heat energy is generated by current flow through a solid-state heating element.

Model 4010 is a square-flange carburetor with mechanical secondaries. These double-pumpers come in 600 and 750 cfm sizes. This one has a manual choke, but electric-choke versions are offered, or the manual ones can be converted with a kit.

Model 4010 with vacuum secondaries: 600 and 750 cfm versions are available. Manual and electric chokes are offered.

A great universal Q-jet replacement: Model 4011 with diaphragm-operated secondaries. Comes in 650 and 850 cfm ratings, mechanical or diaphragm-operated secondaries, electric or manual choke.

Thermocouples to measure exhaust-gas temperatures (EGT) have been installed in these dynamometer headers just downstream of each exhaust valve. Low-voltage signals are sent to an instrument that indicates temperature. A too-lean fuel/air ratio results in high EGT; too rich and EGT will be low. Cylinder-to-cylinder variations should not exceed 100F. J-shape bowl vent tubes identify that this Model 4010 carburetor is for marine use.

Select & Install Your Carburetor

4

IMPROVED BREATHING INCREASES VE AND POWER

Stock passenger-car engines are fairly efficient air pumps up to 5000 rpm or slightly higher. And they provide a reasonably flat torque curve over this operating band. Pumping capabilities can be idealized (optimized) to provide better pumping within a narrower rpm band.

This is done by improving breathing. Reducing restrictions that cause pressure drop allows increased charge density to reach the cylinders. Typical improvements include

• Carburetor changes (higher capacity).
• Intake-tract changes (manifold through the ports and valves).
• Exhaust system (headers or free-breathing mufflers).
• Valve timing and lift (camshaft).

These design changes make a more efficient pump and optimize performance at a *particular rpm.*

But, there's a trade-off! Making the engine a better pump lifts the torque peak and the entire torque curve to a higher rpm band. Lower rpm performance is *reduced* accordingly.

Not all engines can be improved for higher performance (better breathing) without extensive modifications. In many cases the designers purposely optimized low rpm performance. They used:

• Small-venturi carburetor.
• Tiny intake manifold and/or cylinder-head passages and ports.

Ford 427 cid/425 hp original equipment manifold with Holley vintage replacement carburetors. These 550 cfm four-barrels have diaphragm operated secondaries. Primary has electric choke. 0-80431 primary and 0-80432, secondary.

• Small valves actuated by a short-duration, low-lift camshaft with lazy action.
• Low compression.
• Combustion-chamber design.
• Restricted exhaust system.

Truck engines, low-performance passenger-car engines and industrial engines are good examples of these design characteristics.

As you may have heard, "You can't make a race horse out of a mule." Keep such factors in mind when selecting a carburetor. It is difficult—and extremely expensive—to upgrade engine performance where there are built-in restrictions. NASCAR engines with their restrictor plates under the carburetors are perfect examples of the problem!

If the engine is restricted so peak power occurs at 4000 rpm, don't select a carburetor to feed that engine at 6000–7000. *The engine will never run at those speeds.* A too-large carburetor will only worsen the previously available performance.

UNPACKING YOUR CARBURETOR

If you are buying your carburetor in a store, unpack it before you pay for it. Look at the inside of the box. Note whether one side shows damage from carburetor movement. If so, check that side of the carb very carefully. Look it over slowly and thoroughly to make sure there is no shipping damage. Don't hurry your exam-

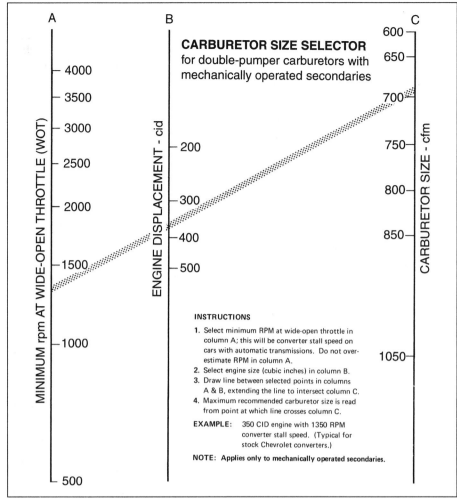

CARBURETOR SIZE SELECTOR
for double-pumper carburetors with
mechanically operated secondaries

A
MINIMUM rpm AT WIDE-OPEN THROTTLE (WOT)
- 4000
- 3500
- 3000
- 2500
- 2000
- 1500
- 1000
- 500

B
ENGINE DISPLACEMENT - cid
- 200
- 300
- 400
- 500

C
CARBURETOR SIZE - cfm
600
650
700
750
800
850
1050

INSTRUCTIONS

1. Select minimum RPM at wide-open throttle in column A; this will be converter stall speed on cars with automatic transmissions. Do not over-estimate RPM in column A.
2. Select engine size (cubic inches) in column B.
3. Draw line between selected points in columns A & B, extending the line to intersect column C.
4. Maximum recommended carburetor size is read from point at which line crosses column C.

EXAMPLE: 350 CID engine with 1350 RPM converter stall speed. (Typical for stock Chevrolet converters.)

NOTE: Applies only to mechanically operated secondaries.

If your car has an automatic transmission, make sure you know the *converter stall speed* before using this chart. If in doubt, use the figure shown for a typical Chevrolet converter (1350 rpm). If using a modified converter for racing, make sure the stall speed is what you think it is.

If your car has a manual transmission, use the lowest rpm at which you use WOT. This must be a very conservative rpm (on the low side) and is found by observing your driving habits. Watch the tachometer! The heavier the vehicle and the lower the numerical axle ratio (higher gear ratio), then the lower this rpm must be.

With engines from 300–400 cid, the right choice usually works out to be a 650 to 700 cfm carburetor. A light car, such as a Camaro or Mustang may be able to use a 700 or 750 cfm unit, especially with a high numerical gear ratio (low gear ratio).

When in doubt, select a *smaller* carburetor size because it will typically give better acceleration times, even though power may fall off slightly at top rpm. You'll be satisfied with a smaller carburetor nearly every time!

Regardless of evidence to the contrary, a lot of carburetor buyers convince themselves that "bigger is better." Holley sells more large carburetors (800 and 850 cfm) because of the faulty logic that if a 650 is good, then an 850 must be that much better. Not true! Co-author Urich regularly gives this advice about airflow capacity: *"Don't buy it if you can't use it!"*

ination. Take your time to look at the carburetor carefully.

Holley packaging engineers constantly improve packaging, but their best efforts can't offset all mishan-

dling. The *outside* of a carton may look perfectly intact while the contents are damaged.

Visually inspect the carburetor. Hold the carburetor as you operate the

MECHANICAL SECONDARY SELECTION

A mechanical-secondary carburetor has an inherent advantage over any controlled-secondary (air valve or diaphragm-operated) carburetor. This assumes the same capacity (airflow). At any engine speed lower than where the controlled secondaries or air valve completely open, the mechanical secondary offers a lower pressure drop at WOT. Lower manifold vacuum creates a denser charge and greater engine output. But great care must be used in selecting mechanical-secondary carburetor size (capacity).

A controlled secondary operates only on the primary side until a specified higher airflow is reached. Thus, primary velocities are high, giving strong metering signals. When a mechanical secondary is "punched" wide open, air flows through all of the venturis. Velocities and signals are low. The pump shot has to supply a "cover-up mixture" until engine rpm provides airflow adequate to start the main system.

The larger the carburetor, the higher the airflow required to start the main systems. With a too-large carburetor, the pump shot is used up before the main systems start. This results in a sag or bog. Double-pumpers help here. They are offered in 50 cfm increments from 600–850 cfm.

Holley Technical Representatives constantly see carburetors that are too large for an engine installed. Mike Urich designed this handy chart to aid in selecting double-pumpers. Follow it to select the right one for your application.

throttle lever to make sure the throttles open fully and close without binding or sticking on both primary and secondary barrels. On vacuum-secondary models hold the primary throttles open and actuate the secondaries manually.

A bent throttle lever can occur. When it does, the throttles may remain partly open (not return fully to idle or fast-idle position) when the lever is actuated. Is it all there? Are any parts missing? The instruction sheet may provide illustrations that will help in your inspection.

If the carburetor doesn't appear to be exactly right, don't buy that one. Ask to see another one and then check it out exactly the same way.

If your carburetor came through the mail, by United Parcel Service or some other delivery method, inspect it immediately when it arrives. Open the box and make sure it is the *correct* carburetor. That *is* important, as the wrong carburetor may have been shipped. Then check it thoroughly *as we just described.*

If there's a problem, proceed carefully if you try to fix it. If it is simple–such as a bent throttle lever–you

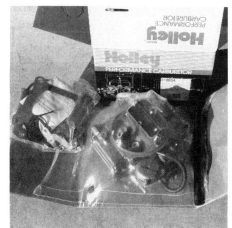

Holley's packaging engineers constantly improve packaging so their products will be in good condition when they arrive. Heat-shrink plastic material secures this 0-9834 carburetor to a corrugated-cardboard "pallet" inside the carton. Plastic "hat" on top of the airhorn keeps bowl vents from tearing through the plastic.

may be able to remedy the problem easier than returning the unit. However, if the casting has been damaged or if a shaft is bent, the carburetor must be replaced.

The firm you bought the carburetor from will replace it. Find out what the supplier requires *before sending it back for replacement.* In some instances,

damage claims must be settled between yourself and the freight company (called the *carrier*). In other cases, the supplier takes care of this for you. Try to get the supplier to handle any shipping-damage problem. *Do not send the carburetor to Holley.* Warranty claims are only handled by the dealer.

Timely Ordering–Order the carburetor *early.* It may not be obvious, but one of the first things to remember in building your engine is to get the correct parts together. Then assemble and test the car before setting impossible dates for completion and first competition attempts. Many cars don't show up at major events because the builder was overly optimistic about the amount of time required. The owner loses an expensive entry fee and misses the race.

It is possible the supplier may not have the unit you need in stock. He could have just sold the last one and be waiting for a new shipment to arrive. Or, the unit you require may be so popular that the factory temporarily cannot supply enough carburetors. Another possibility is that the dealer *never* stocks the carburetor you want, so he will have to special order it for you.

COMPARING CARBURETOR FLOW CAPACITY
When comparing carburetor flow capacity, it is important to know that one- and two-barrel carburetors are measured on a different basis than four-barrels and fuel-injection throttle bodies. Flow capacity of one- and two-barrel carburetors is measured at 3.0-in.Hg pressure drop. This figure has been used for many years because this is a typical WOT pressure drop across the carburetor (manifold vacuum) in low- and medium-performance engines. Four-barrel carburetors are measured at a pressure drop of 1.5 in.Hg. To relate the two measurements, use the formula:

$$\frac{\text{cfm flow at 3 in.Hg}}{1.1414} = \text{Equivalent flow at 1.5 in.Hg}$$

EXAMPLE: For a 500 cfm two-barrel
$$\frac{500}{1.414} = 354 \text{ cfm four-barrel equivalent}$$

Holley standardized flow rates at 1.5-in.Hg pressure drop for three- and four-barrel carburetors and fuel-injection throttle bodies; 3.0 in.Hg for one- and two-barrel carburetors.

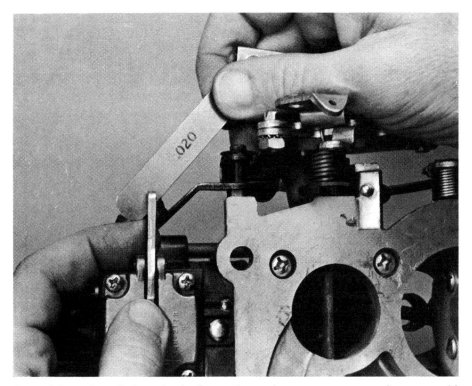

Essential pre-installation checks for each accelerator pump on carburetors with removable fuel bowls: (1) no clearance between the actuating-lever screw and the diaphragm lever at curb idle, (2) at WOT you should be able to move the diaphragm lever to allow inserting a 0.015—0.020-in. feeler gauge as shown.

So, order your carburetor early—ahead of when you absolutely must have it. If you wait until the day before you have to race or use the car, you could end up buying the wrong carburetor. This means you will buy the right one later on, or use a carburetor not precisely right for your application.

Get the other pieces you need at the same time: fuel lines, fuel pump, spacer or adapter, studs and nuts, extra gaskets and jets, air cleaner, tubing wrench, new fittings and a fuel filter.

BEFORE INSTALLING YOUR CARBURETOR

Inspection—You already made an inspection when you got the carburetor, but repeat that again. Make sure the carburetor operates easily and opens to full throttle on both primary and secondary barrels. It should return easily to idle. Throttle shafts are equipped with adjustable stops. Throttle levers should return to rest against their adjustment screws when the choke is open, *not against the throttle bores.*

Screw Check—Check for loose screws attaching the top of the carburetor or float bowls and screws holding the accelerator-pump shooter/s. Also check the screws holding the castings together. Gaskets compress after they've been installed. It is important to check all screws so there will not be any fuel or air leakage when the carburetor is installed. Don't use too much force—anyone can strip threads by applying too much torque to the screws.

This is a good time to install power-valve blow out protection if the carburetor doesn't have it. (See page 55).

Hose Fittings—When you unpack the carburetor you will notice that some or all of the hose fittings are closed with rubber shipping plugs. These can be left installed when no hose will be attached to that fitting.

If the 3/8-in. PCV fitting extending from the rear of the base isn't used, replace it with a shipping plug or a piece of hose plugged at one end. Either must be clamped onto the fitting so a backfire won't blow it off. Or, close the tube with epoxy.

Don't use rubber compounds like Silastic RTV to seal this fitting because gasoline turns them to jelly.

Disassembly for Checking—If you plan to take the carb apart to look at its insides, expect to replace bowl and metering block gaskets because these usually self-destruct on disassembly. Put the carb back together immediately upon finishing your inspection.

You will need new gaskets, spray-on gasket remover and a lot of patience. This is not a 5-minute job. Expect to spend at least an hour carefully removing the old gaskets from the metering block/s and fuel bowl/s.

Accelerator Pump Check—Check for correct accelerator pump adjustment. Open the throttle wide, then move the diaphragm lever by hand. This 2-1/8-in. long lever pivots in the pump-cover casting. The lever must have an additional 0.015–0.020-in. travel at WOT throttle so the pump doesn't bottom in its housing. This helps to ensure the throttle won't stick open because of excessive friction of the pump-operating lever against the plastic cam. See photo above.

Also check that the pump-operating lever contacts the plastic cam at one end and the diaphragm lever at the other at idle. You should be able to wiggle the diaphragm lever back and forth slightly (horizontally) with your fingers. The slightest throttle movement should be transmitted through the pump lever, causing the diaphragm lever to move.

The only exception: When a green cam is used, check for 0.010-in. clearance at WOT.

Run Carb Before Changing It—In general, bolt on the carburetor and run it *before making any changes!* Chances are the carburetor will be very close to correct. Start with what Holley engineers have found to work successfully. They produce thousands of carburetors every year–for all kinds of applications.

REMOVING YOUR CARBURETOR

Have a notebook and pencil handy for making sketches or notes of what you are taking apart. This is the mark of a professional. Keep track of what has happened so you won't have to guess on reassembly. It's a good idea to have a fire extinguisher handy. Extinguish any pilot lights in the area or move to another place where there are no pilot lights.

Air Cleaner–Remove the air cleaner, carefully detaching any vacuum lines to the cleaner and marking them with masking tape so they can be reassembled correctly. Cover the top of the carburetor with a rag or masking tape. You don't want spare nuts or other pieces falling into it.

Recommendation: Before proceeding, tag and number all vacuum hoses and other lines. This will help during installation. Remove the existing carburetor as follows but be aware that not all items apply to every application.

Fuel Line–Remove steel fuel-line fitting carefully because you'll probably reuse it. A fuel-line wrench that contacts four nut flats should be used. These are called *flare-nut* or *tubing wrenches.*

Fuel-line nuts are soft and tend to round off instead of turning. When this happens you will have to wreck the nut by using Vise-Grip pliers to turn it. To replace the nut requires cutting the fuel line very close to the end of the tubing. Cutting and reflaring will require a tubing cutter and a flaring tool. Rent them from an auto-parts house or hardware store.

Getting the line off the engine to allow this work requires disconnecting the fuel line at the fuel pump. If the nut at the carburetor rounded off, the same

Throttle and transmission linkages, smog equipment connections and so forth are complex. Make a sketch BEFORE you take the manifold and/or carburetor off. Tag lines and connections with tape to identify each. Use any tactic necessary to help your memory: photos, drawings or sketches.

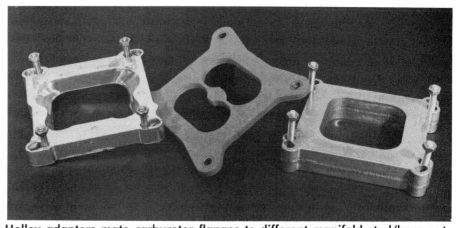

Holley adapters mate carburetor flanges to different manifold stud/bore patterns. Adapters must be used in some instances. Avoid flow losses by using a manifold with the matching stud/bore pattern for the carburetor. Holley adapter 17-6 at left puts spread-bore onto square-flange manifold or vice versa. Center heat insulator gasket, Holley 108-18, is 5/16-in. thick with phenolic eyelets. Adapter at right, Holley 17-27, adapts square-flange carbs to spread-bore manifolds or serves as a 1 in. spacer for square-flange carbs.

wrench is sure to round off the one at the fuel pump. So, buy a tubing wrench before taking off the fuel-pump fitting or you may have to create an entire new fuel line. Using the correct wrench is cheaper in the long run.

Seal the open end of the fuel line with masking tape or a clean, lint-free cloth so no foreign particles can enter.

Unless you plug the end of the fuel line so it can't leak, do not let the end of the fuel line droop below the fuel-tank level or you'll have gasoline all over your garage floor. You did remember to turn off all pilot lights in

the area, didn't you? You don't want a fire to turn you and your vehicle into crispy critters.

Throttle Linkage–Disconnect throttle linkage, automatic transmission controls and cruise controls from throttle lever. Save any retaining clips. Take off throttle-return spring/s and note anchor points for correct reassembly.

PCV Hose–Disconnect PCV hose, if it's attached to carburetor.

Vacuum Lines–Disconnect distributor spark hose/s, labeling these to keep vacuum and timed-spark hoses separate

STACKING GASKETS

Avoid gasket stacks or packs. These compress unevenly and the carburetor base can warp, binding the throttle shaft. They can cause a corner of the throttle base (mounting flange) to break off. Use the gasket supplied with the carburetor. Always use a thin gasket such as the 0.025-in. one supplied by Holley with a diaphragm-secondary carburetor. No gasket sealer is required if the carburetor base and manifold flange are clean and flat.

Don't stack gaskets to get clearance for lever operation, regardless of how badly you want to get your engine running. Do it right or you could end up buying an entire new carburetor body or throttle base.

These parts are not ordinarily stocked by even the largest dealers, and there is often a long wait for pieces to reassemble your carburetor. If the throttle body is not a separate part, you will have to buy a new carburetor because the main body is not sold as a replacement part.

If you need additional clearance to install a carburetor with a high-capacity accelerator pump, make a 1/4-in.-thick aluminum spacer and use a thin gasket on top of the spacer and one on the bottom. Spacers are available from Holley, too.

Heat shield surface marked TOP goes against carburetor base. This is Holley 108-20 GM 3884576 (baffle). Center gasket, Holley 108-15 or GM3884574, fits between heat shield and manifold flange. Arrowed sections must mate or there will be an exhaust leak. This arrangement was used on Chevrolets through 1969. In 1970 the heat shield was eliminated. Gasket/spacer at right, Holley 108-25 or GM 3998912 has bushings around stud holes.

for correct reinstallation. Disconnect any other manifold vacuum hoses connected to the carburetor.

Other Hoses—Disconnect fresh-air hose and label it. Remove and label any other hoses such as gulp-valve hose (for air-injection-equipped cars) and connections to power brakes and so forth. Keep dirt out of all the hoses and lines you've removed.

Choke Linkage—Detach choke linkage at carburetor, noting whether it pulls or pushes to close the choke plate.

Remove Fasteners—Remove nuts and/or bolts and lockwashers attaching carburetor to manifold. Move any cable brackets out of the way. If you drop any nut, lockwasher, cotter key,

piece of linkage or whatever, STOP! Find and recover it *before* proceeding.

Observe and note the position of any brackets held on the engine by the carburetor-attachment hardware. Lay these aside so they will not fall into the manifold when the carburetor is pulled off. Before you lift the carburetor, check carefully. Make sure nothing loose can fall into the manifold . . . such as a nut, bolt, fitting, cotter key or whatever.

Lift Carburetor—Carefully pull carburetor off manifold. If it sticks, tap each side gently with a rubber or plastic mallet. Stuff a rag into the intake manifold opening/s. Clean the intake-manifold mounting pad thoroughly. Take care to keep gasket pieces out of the manifold.

INSTALLING YOUR CARBURETOR

Install your new carburetor as follows, but be aware that not all steps apply to every application and installation.

New Studs—Install any new studs. Use the jammed double-nut technique to screw in the studs with a wrench. Use Loctite on studs.

Manifold Gasket—Install manifold-flange gasket. If a metal heat shield protects the carburetor base from exhaust gases in the heat-riser passage, place a thick (0.080-in.) manifold-flange gasket under the steel heat shield. No gasket is used between the shield and the carburetor. Pay attention to the gasket combination used by the automobile manufacturer. Duplicate it as nearly as possible.

Idle Solenoid—Sometimes the instructions ask you to transfer the idle solenoid and other items from the original carburetor to the new one. These may include linkage ball connections or lever extensions. If the original throttle lever has a non-removable ball, a corresponding ball will be in the loose parts supplied

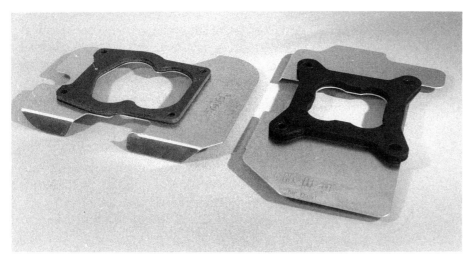

Reduce the heat on your carburetor (and inlet charge) by using one of these Holley heat shields under the carburetor. Spread-bores use Holley 108-69 at left. Any square-flange except 4500s use Holley 108-70 at right. Each includes an insulator gasket.

with the carburetor. These are in a plastic bag.

Adapting to the Installation—You may have to remove some items from the new carburetor to adapt it to a particular application. Examples include a Chrysler-type throttle-lever extension on some two-barrel and four-barrel carburetors with universal throttle linkage. This must be removed to use the carburetor on Chevrolet or Ford installations.

Carburetor Base Gasket—One of the easiest ways to prevent manifold leaks is to check the carburetor base gasket. Hold it against the manifold flange. There must be adequate clamping area to hold the gasket. Make sure all openings match up. If the gasket is correct for the manifold, hold it against the carburetor base to see if it fits the throttle bores, mounting holes and base shape. Be especially careful when installing a 4010 carburetor on an Edelbrock manifold with a narrow mounting pad. Place gasket on manifold.

Carb on Manifold—Place carburetor on manifold and add any brackets held in place by the attachment hardware. These may include throttle linkage, transmission kickdown, cruise control and solenoid brackets if such are used. Start the fuel line nut/s into

the carburetor carefully. Tighten them. Install nuts and/or bolts and tighten them gently against the carburetor flange. Then cross-tighten the bolts/nuts alternately to 5–7 ft-lb torque. This is not much force!

Overtightening can warp the throttle body, binding the throttle shafts so throttle action is stiff. In the worst case, overtightening may snap off a corner of the base. That *is* expensive and time-consuming.

Vacuum-operated secondary throttles may never open if the base is warped by overtightening during installation. Hold the throttle wide open and operate the secondaries by hand. Make sure they open easily against the spring and close freely by spring action. If you fail to check this point, the secondaries may never work and you'll never get the performance you paid for.

Linkages—Connect throttle and transmission linkage. Attach throttle-return spring. Operate linkage to ensure nothing binds as the throttle is fully opened and then closed. With the choke held wide open, make sure the throttle returns to idle without sticking or binding.

Make sure the levers don't hit anything when the throttle linkage is actuated from idle to WOT. Have a friend actuate the linkage with the

foot pedal as you observe the linkage. See for yourself whether the throttle plates are 100% open in both the primary and secondary barrels.

Don't try to make this check by operating the throttle with your hand at the carburetor or by looking at what the levers *appear* to be doing. Slack or play in the linkage may allow full throttle when moved at the carburetor, but give less than full throttle when the foot pedal is depressed.

If foot-pedal movement doesn't give full throttle, examine the linkage to see what minor changes or adjustments may be needed in the linkage connecting to the carburetor. Don't change the linkage or levers on the carburetor itself. Simple adjustments are usually provided on the vehicle cable housing or linkage rods.

Don't think checking for full throttle is only for amateurs. Experts have been tripped up on this point—whether they admit it or not.

Check both the pump and the throttle levers to make sure nothing is in the way. Sometimes you'll have to install a spacer to clear high-capacity accelerator pump/s. Install the air cleaner and make sure it doesn't hit any portion of the linkage. Operate hand choke to check whether any clearance problem exists.

WARNING: Any binding or interference could cause the throttle to stick during operation and could result in a loss of throttle control (uncontrolled engine speed). This check should be made by operating the accelerator pedal from inside the vehicle.

Choke Linkage—Install choke linkage for divorced choke. Hold throttle lever partially open and activate the choke manually to make sure it operates freely. If the choke is electrical, make all the necessary connections. If the choke is integral, attach the necessary lines from the manifold.

Hoses—Reconnect appropriate hoses to carburetor. Transfer any additional fittings from the old carburetor if needed. Temporarily plug vacuum hose to air cleaner. Any vacuum connections on the carburetor which are not used should be closed with rubber plugs or caps.

Fuel Line—Flush fuel line to ensure no foreign particles enter the carburetor. Disconnect primary wire to coil (it's the small one connected to the + terminal) and insulate its end with a piece of tape so it won't spark. Hold a can under the open end of the fuel line and catch the fuel as you crank the engine several revolutions with the starter. Discard this fuel safely.

Install an in-line fuel filter to keep dirt out of your new carburetor.

Connect fuel line to carburetor or fuel-line assembly. In some instances a minor bend in the fuel line may be needed. Or, you may have to shorten the fuel line and reflare the end.

Air-Cleaner Stud—Remove air-cleaner stud from old carburetor and install it in the new one, or use one provided with the carburetor. If the new manifold (and carburetor) is taller than the old one, the air cleaner or stud could dent in hood. Check this clearance carefully before closing the hood for the first time.

Check for Fuel Leaks—Set the parking brake and block the drive wheels to prevent any vehicle movement. Crank the engine or turn on the electric fuel pump to fill the carburetor with fuel. Check the fuel-inlet fitting/s for leaks. Tighten fittings to eliminate any leaks.

First Start—Reconnect the coil's primary wire, make sure the vehicle is in Neutral or Park, and ensure that it's safe to start the engine. Block the drive wheels and make sure the area in front and behind the vehicle is clear.

Now, start the engine. Depress the accelerator pedal to floor and allow it to return. This charges the manifold with fuel and sets the choke and fast idle. With foot off accelerator pedal, crank engine until it fires. Repeat the previous two steps as required.

Flooded?—If carburetor floods over, there may be dirt in the needle-and-seat assembly.

If flooding continues, stop to correct cause. First try tapping the carburetor lightly with a screwdriver handle. If flooding doesn't stop, remove the fuel line and take off the carburetor top or float bowl.

Check float setting. If it's OK, chances are there is dirt under the needle. If the needle and seat are separate items pull the needle out of the seat and blow the seat with compressed air from both sides. If the needle and seat assembly is not a take-apart type, unscrew the entire needle and seat from the float bowl and agitate the assembly in solvent, then blow it dry with compressed air.

Setting Fast-Idle—Read the sidebar on page 83 for precautions to take when setting fast-idle speed. This is usually adequate for most vehicles. If a particular application requires adjustment, use the following procedure. With a fully warmed engine running, advance throttle and place fast-idle cam so screw contacts top step of cam. On this step, set fast idle at approximately 1700–1900 rpm.

If your vehicle has T.C.S. (Transmission Controlled Spark advance) fast-idle speed adjustment is made with no vacuum advance and in Neutral. When the engine is cold, there will be vacuum advance and the speed may be too fast. Make changes to get it where it suits you.

Setting Idle Speed—Again, read the sidebar above for precautions to take when setting idle speed. Set idle speed according to the decal in the engine compartment. On automatic-transmission-equipped cars the idle speed is usually set *with the parking brake on* and the transmission in Drive. On manual transmissions, put in Neutral and set the parking brake. Start the engine and check for vacuum and/or fuel leaks. Correct as necessary–temporarily plug vacuum hose to air cleaner.

Don't remove the idle-limiter caps. Idle-mixture screws are factory-preset for proper emission values. Engine differences may necessitate slight adjustments to obtain a smooth idle.

NOTE: It's easier to make idle settings with the air cleaner removed, but to ensure final readings are correct, they should be taken with the air cleaner installed. If there is a throttle solenoid, set curb-idle speed by adjusting nut on the end of the solenoid (solenoid energized). Set the kill idle speed by adjusting the idle-speed screw on the carburetor with the solenoid de-energized. See underhood decal for idle rpm and mixture setting.

Set Timing & Dwell–Before making final adjustments on the carburetor, check the ignition timing and dwell. Reset if required. Disconnect and plug the distributor vacuum line

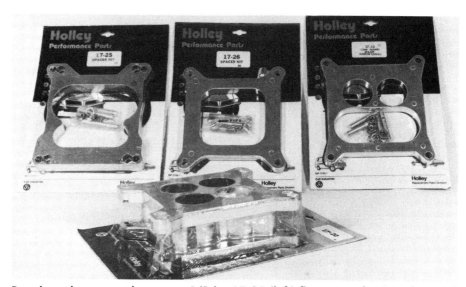

Popular adapters and spacers: 1/2 in. 17-25 (left) fits square-bore and spread-bore patterns, adapting between the two. 17-26 (middle) is a 1/2 in. spacer for square-bore pattern carburetors and manifold flanges. 17-15 (right) is required for 460 cid Fords when adapting Model 4150/60/80 to Ford Model 4300D manifolds. 17-20 (front) is 2-in. spacer to raise Model 4150/60/80 carbs above manifold flange.

during this operation and reconnect the hose when completed. With the engine at normal operating temperature, place the transmission in Drive and adjust the hot-idle speed to the rpm indicated on the decal.

Air Cleaner–Install the air cleaner stud (provided). Install the air cleaner gasket (provided). Mark the stud 1/2 in. above the air cleaner. Remove stud and cut any excess with a hacksaw. Reinstall the stud with Loctite. Connect the manifold vacuum to the

air cleaner connection. Connect any other air cleaner hoses. Check hood clearance to stud and air cleaner before closing hood to avoid expensive dents in your hood.

GET ACQUAINTED
Once you have installed your carburetor, live with the new combination for a few days before beginning any tuning efforts. Naturally you will want to set the idle correctly (to exact specifications if your car has emission-control

devices), check the ignition timing with a timing light and make sure the spark plugs, points, cap and wires are all in tip-top shape.

By driving the car, you will learn what you have–a baseline condition. If you are drag racing, get your times consistent by working on your driving technique *before* you start tuning. There are enough variables without adding inconsistent driving. This is especially important if you have changed from an air-valve-secondary carburetor (such as a Q-Jet) or a vacuum-operated-secondary carburetor, to a double-pumper with a mechanical-secondary linkage.

More tuning details are provided in Chapter 8, *Performance Tuning.*

CHECKING FUEL LEVELS OR FLOAT SETTINGS

Installing the carburetor and pumping fuel to it is the only easy way to check whether the needle/seat will seal correctly. Turning the carburetor upside down and blowing into the fuel inlet is not an adequate test. Your lungs cannot develop fuel pump pressure.

When the carburetor is installed and the engine is started, see if there is any flooding into the manifold. If there is no seepage or flooding into the throttle bores, chances are the fuel level is OK. If the carburetor doesn't have externally adjustable inlet valves and sight plugs, test drive the car. If it drives normally without dying at stops, the fuel level is probably fine. If the fuel level is too high, fuel spilling into the throttle bore/s will cause the engine to die at stops or after severe braking.

High-performance Holley carburetors have the advantage of externally adjustable inlet valves. Fuel-level sight plugs allow checking and setting fuel levels without taking the carburetor apart or taking it off of the engine.

Setting Fuel Level–If the engine has a mechanical fuel pump, turn off the engine. Lower the fuel level by loosening one bowl screw and allow-

As shown here, combination metering plate and bowl gasket swells into the secondary metering-plate channels, reducing their capacity. This brand-new, never-run 0-1850 carburetor's gasket raised up into these channels. Contact with gasoline increases this swelling and may affect performance over time. This can be eliminated by using a thin steel plate sandwiched between the big gasket and the metering plate, shown here with new gaskets, lower right. Some 4160, 4175 or 4180 carburetors have this thin steel plate (see 4150/4160 exploded view, item 98 on page 166). If not, add Metering Body Plate 1034-1993 against the large gasket and a paper gasket such as 108-27 between the steel plate and the secondary metering plate. Thanks to Frank Townsend for this tip on an inexpensive improvement.

ing some fuel to leak out into a container (soda can cut in half) or a shop rag. Retighten the screw. Then run the engine to see whether the fuel level returns and stays where it is supposed to be.

Adjust the inlet valve so the level is just below the point where it would

spill through the sight-plug hole. Check to see it stays there. Then raise the level until the fuel just spills through the plug hole. That way, you know the inlet valve is holding and the level is correct.

With an electrical pump, turn on the pump without running the engine.

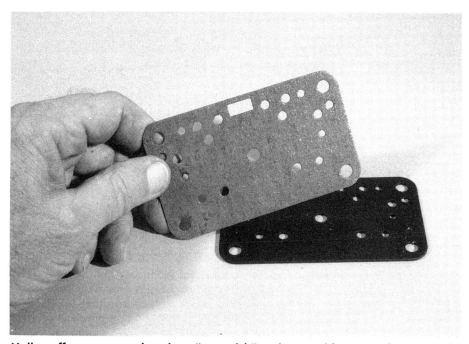

Holley offers some gaskets in a "non-stick" variety to aid tuners who must take their carburetors apart for jet changes and adjustments. These "-1" gaskets are light-tan, as compared to the factory-installed black gaskets that are guaranteed to stick and self-destruct on disassembly. Applying talcum powder or spraying a new gasket of either type with WD-40 or silicone spray before assembly usually allows taking the carburetor apart several times without hurting the gasket.

Make sure the inlet valves are holding and not "creeping." Then run the engine to make sure the level stays where it should. Be sure to reinstall the sight plugs after checking levels.
WARNING: Don't drive the car without reinstalling the sight plugs!

The main problems with fuel-level changes are caused by rough handling during shipping. Jarring forces can bend the tabs that establish the float/needle relationship. We've even seen float-bumper springs dislodged from the underside of floats.

Once a carburetor is installed on a car, the fuel level established by the float setting seldom changes. This means the simple task of level setting is usually a one-time operation with occasional checking to make sure it is still OK.

Don't raise or lower the fuel level except to get it to specification. It has been factory-calibrated for cornering, turns, spin outs, brake stops and accelerations. **Float levels are most critical with high G-loadings because those affect fuel handling.**

Slosh problems in the fuel bowls are severe. Just imagine trying to hold a cup of coffee in a car without sloshing during a burnout or a panic stop! That should give you some idea of what happens inside the carburetor fuel bowl/s. The only time the float level might be changed would be for drags or super-speedway use. The secondary level might be raised slightly, but expect brake-stop stalls as the result.

When a carburetor is being assembled and disassembled for calibration purposes, it is common practice to apply talcum powder to both sides of the gasket so sticking won't occur. Then use a new gasket when you are ready to reassemble the carburetor permanently. That will prevent dangerous fuel leaks.

BLUEPRINTING HOLLEY CARBURETORS

"Blueprinting" has become an important word in the enthusiast's vocabulary. As applied to an engine, it means correctly relating all parts of a production engine by matching and mating, balancing and rematching. The goal is to get the combination into a near-perfect state suitable for performance application. Blueprinting is often done on brand-new engines. This attention to detail can provide impressive performance improvement as an immediate payoff for the time and care invested in the project.

Blueprinting is now applied to the precise preparation and rebuilding of any engine or chassis component. But, it is a mistake to use the word or concept in relation to a Holley high-performance, original-equipment or production-replacement carburetor. Tearing a new carburetor apart and attacking it with drills, trick linkages and so forth is like overhauling a new watch or micrometer when you first get it–something you probably wouldn't consider doing. There is too much chance of ruining it and voiding the guarantee.

Carburetors are precisely made. They must accurately meter fuel and air into the engine. *They have to be correct before they are shipped.* Every carburetor to be used as original or replacement equipment on a production automobile must pass computer-controlled tests that check whether that individual carburetor performs. These tests ensure it will provide correct performance and low emissions when it is installed. If the carburetor is "out of spec," any "blueprinting" must be done by Holley before it is shipped. This requirement is law.

Production techniques have been developed over years of carburetor making. Each unit must work right when an automobile manufacturer, or an automotive enthusiast, bolts one onto an engine. Regardless of what you may have read to the contrary, very little needs to be done, or can be done, to one of these precise devices to "make it better" before you use it. Trick changes or special assembly techniques are not required to make the carburetor right for high performance. These would be incorporated and fully tested before the design was released to one of the Holley manufacturing plants.

Numerous magazine articles every year point out "fixes" to be made to new or used carburetors for more performance. Almost without exception, these have been tried at Holley's engineering labs and in field tests. And they are discarded as being unworkable or not generally applicable. Using care in selecting, installing and tuning your carburetor is the best "trick" to get the utmost performance and satisfaction from it.

WHICH CARBURETOR IS RIGHT FOR YOUR ENGINE?

Enough different Holley carburetors are available as aftermarket models so almost any engine can be mated with a correct carburetor for the engine and application.

Some builders fall in love with a certain Holley size, such as the 850 cfm and disregard all other sizes. A larger or smaller carburetor might be correct for a particular application. Different types of carburetors are built because there are different kinds and sizes of engines. Their requirements are not the same. Holley would certainly prefer to build fewer types if this could be done and still match the needs of the engines that will use them.

It's obvious from outside appearance that many carburetors look quite similar. But, the mechanical secondary, vacuum secondary, double-pumper, single-pumper, non-pullover discharge nozzle with check valve just under it, rubber-type pump-inlet valve, ball pump-inlet valve, side-hung and center-hung float, and so on are all variations born of specific needs.

This may not be obvious if you are not a manufacturer, but every change costs money to design, to build and supply. So, these pieces exist for a reason. After you have read this book, the reasons will be easier to under-

---

SORTING OUT GASKETS
Throttle bores of square-flange Holley four-barrels range from 1-3/16—1-3/4 in. Spread-bore 4011 and 4165/75s have 1-3/8-in. primary and 2-in. secondary bores.

Always check a new gasket against the carburetor base. Make sure it doesn't hang into the throttle bores to act as a restrictor or obstruct throttle operation. Some GM manifolds have heat tracks to warm the carburetor base with exhaust gas (page 80). The following info from the Chevrolet Parts Book and Holley's Performance Catalog may help with your application.

Square-Flange Holleys—To mount on manifolds with heat tracks (through 1969), use Holley 108-10 (1/16-in. thick) with GM 3884575 0.010-in. stainless-steel heat shield.

For manifolds without a heat track, use Holley 108-10 or a thin gasket (0.020-in.) for vacuum secondary carburetors. A thin gasket won't compress much, so there's little chance of warping the base when tightening the flange nuts. Holley's 5/16-in.-thick 108-14 can be used with mechanical secondary carburetors if you tighten carefully so as not to warp the carburetor base.

Holley (Non-Spread-Bore)—For 1970 and later small blocks use Holley 108-12, a 5/16-in./-thick gasket with phenolic bushings around the stud holes. Or, install Holley 108-51 or -53 0.260-in.-thick heat-insulator/spacer gasket.

Holley 4011 or 4165/75—If replacing Quadrajet through 1969 models, use Holley 108-20 0.010-in. stainless-steel heat shield or GM 3884576 (baffle) against carburetor base. Holley 108-15 or GM 3884574 between heat shield and manifold flange.

If replacing Q-jet on 1970 and later models, use Holley 108-25, a gasket with phenolic bushings around stud openings. Or install GM 3998912.

Aluminum Heat Shields—For Holleys (any square-flange except 4500s), use Holley 108-70 or GM 3969835. These are 7 X 13 in., 7/16-in. thick with insulator gasket. Added 1/4-in. spacer (Holley 108-51, GM 3999198) required or cutout to clear 50cc accelerator pumps.

For spread-bores use Holley 108-69 or GM 3969837. These are 9 X 10 in., 5/16-in. thick with insulator gasket. May require cutout to clear rear 50cc accelerator pump on 4165s, or use 1/4-in. spacer (Holley 108-25 or GM 3998912).

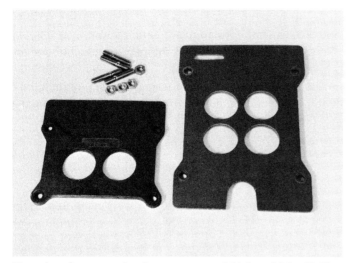

Heat-insulator gasket/spacers are 1/4 in. thick. Holley 108-52 (left) fits Model 2300. 108-51 (right) fits Model 4150/60 with 1-9/16-in. throttle bores; 108-53 (not shown) fits 1-1/2-in. bores. Gasket/spacers reduce carburetor temperature, and improve hot starting and performance.

Have the correct gaskets before taking apart your 2300, 4150/60, 4165/75 or 4500. Holley's black gaskets self-destruct when you take apart your carburetor. Holley's tan "non-stick" gaskets are not necessarily non-stick. They may self-destruct on disassembly.

stand. You'll be right at home when you specify a particular carburetor for your own or a friend's engine.

Fuel Supply System—Read that chapter for important details on fuel lines, fuel filters and fuel pumps.

Adapters—A carburetor should be installed directly onto its manifold without an adapter. Adapters don't provide optimum airflow characteristics. Many reduce the carburetor's flow capacity. It's best to buy a carburetor that fits the manifold to be used. It is sometimes possible to drill and tap the manifold to fit a slightly different stud pattern.

In the case of a four-barrel manifold with "standard" or "square" pattern, the double-pumper or a vacuum-secondary carburetor should be used. If the manifold is designed for a spread-bore pattern, the 4165, 4175 or 4011 can be used. An adapter may raise the carburetor so a special air cleaner or a hood "bump" will be required to gain hood clearance. And, an adapter can create problems with choke linkage, fuel-line attachment

and fuel mixture distribution.

Heat Insulators/Spacers—Anytime you can insulate the carburetor from heat, do it! Holley has several insulator/spacers to consider. Holley 108-37 is an 11/16-in.-thick phenolic spacer for spread-bores. 108-51 and -53 are for Model 4150/60. 108-52 is for Model 2300.

Spacers—These are often used to raise the carburetor on the manifold for better mixture distribution and improved power output. A spacer or heat insulator may raise the carburetor so a special air cleaner or a hood "bump" will be required to gain hood clearance. Throttle, choke linkage and fuel-line attachment must also be considered.

GASKETS FOR MODELS 2300, 4150/4160, 4165/4175 & 4500

At this writing 14 block, metering-plate and bowl gaskets are used for these carburetors. Few are interchangeable. It's best to get a new set before disassembling your carburetor.

The *Holley Performance Parts Catalog* lists gasket numbers for most performance carburetors. Renew Repair Kits may contain more than one type of gasket so that kit can be used for servicing several carburetors.

Holley manufactures two types of gaskets. Black ones are adhesive-treated for best sealing. They make it hard to get the carburetor apart and they can be counted on to self-destruct on disassembly. Tan ones are supposed to be "non-stick," though we've found that they may also destruct on taking the carburetor apart. Always save enough of the gasket so you can compare it with the new one to make sure the replacement is the correct one.

You may have to use STP Carb Spray Cleaner, CRC Gasket Remover or Permatex Gasket Remover to get off pieces of gasket. It takes time and care. Check the instructions on the can. Be extremely careful to keep gasket residue out of the metering-block, carburetor-body openings and your eyes. Wash the parts after cleaning and blow out with compressed air.

The easiest way to get gasket residue off is to let the parts soak in lacquer thinner for a couple of hours. Then the gasket pieces can usually be brushed off easily.

If you are tuning and have to take the carburetor bowls off to access the main jets, coat the gaskets with WD-40 or silicone spray before assembly. In our experience we've found that gaskets treated this way don't leak and can be reused several times,

To make gasket selection less confusing, available gaskets are shown here with information on their usage. Locator pins on the metering blocks help to align the gasket and metering-block or fuel bowl openings. If a gasket looks like it doesn't fit, flop it over and try the other side.

GASKET ASSORTMENTS

These assortments cover popular Holley four-barrels. Each pack contains the number included (in parentheses) and gasket type.

Pack	Contents
108-200	(2) 108-29
	(2) 108-33
108-201-1	Same as above, except non-stick -1
108-201	(1) 108-29
	(1) 108-33
	(1) 108-30
	(1) 108-27
108-201-1	Same as above, except non-stick -1
108-202	(1) 108-33
	(1) 108-31
	(1) 108-30
	(1) 108-27
108-202-1	Same as above, except non-stick -1
108-203	(2) 108-33
	(2) 108-31
108-204	(1) 108-32
	(1) 108-31
	(1) 108-29
	(1) 108-27

(1) 108-13—Secondary metering-plate gasket for Model 4160 Chrysler and outboard Model 2300 on some 3 x 2 applications with diaphragm-operated throttles.

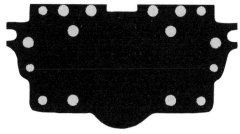

(2) 108-27—Secondary metering-plate gasket for some Model 4160s. Same pattern used on metal 1034-1993 Metering Body Plate.

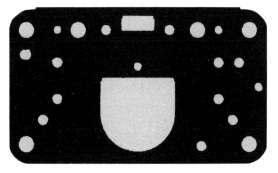

(3) 108-28—Primary metering-block gasket for Model 4160 Chrysler applciations beginning in 1968.

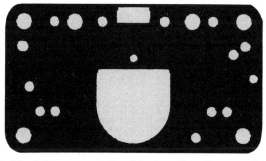

(4) 108-29—Primary metering-block gasket for most Model 4150s, some 4160s, early 4165s and most 2300s. Secondary metering-block gasket on double-pumpers. Not used with accelerator-pump transfer tube. Used for 4500s without inter-mediate idle system. Similar gasket with idle-channel slots discontinued because improvements in gasket material reduced swelling typical with older gaskets.

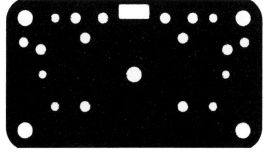

(5) 108-30—Used as a secondary bowl and metering-plate gasket on many 4160s and 4175s.

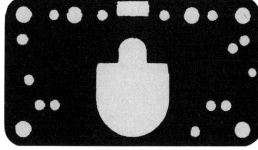

(6) 108-31—Used on same carburetors as 108-29 when equipped with accelerator-pump transfer tube. Used as primary and secondary metering-block gaskets on 4165/75 and a few 4150 carburetors. Used on primary side of some 4160s. Not interchangeable with 108-29, but a -31 can be made from a -29 by cutting a half circle to clear the transfer tube.

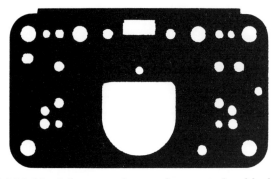

(7) 108-36—Primary and secondary metering-block gasket for Model 4500s with intermediate systems, such as List 6214 and 6464.

(8) 108-55—Primary metering-block gasket for Model 4180.

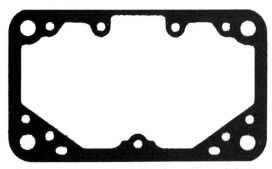

(9) 108-32—Fuel-bowl gasket for all Model 4165 and some 4150, 4160 (primary side) and 2300. Primary-bowl gasket for 4175, except computer-controlled. Accelerator-pump discharge slot differs from 108-33 to work with pump-discharge check in metering block.

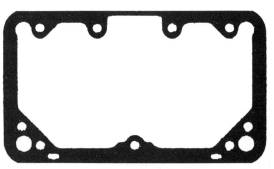

(10) 108-33—Fuel-bowl gasket for Models 2300, 4150/60 and 4500 carburetors. Note: Both 108-32 and 108-33 have only one accelerator-pump hole. Early versions had two holes. Gaskets must be installed so holes line up or there will be no pump shot.

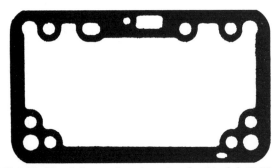

(11) 108-56—Primary-bowl gasket for Model 4180.

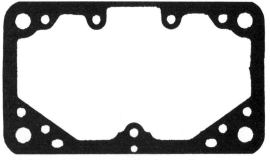

(12) 34-202—Primary-bowl gasket for computer-controlled Model 4175. Primary-bowl gasket for 4150/4160 converted to Mile Dial configuration. Primary- and secondary-bowl gasket for 4150 converted to Quarter Mile Dial.

NOT SHOWN:
108-34—Metering-block gasket for Model 4160, 0-6270-1 for 340 Chrysler.
108-35—Metering-block gasket for Model 2300, List 6425, 650 cfm two-barrel.

Manifold Relationships 5

THE INTAKE MANIFOLD

The intake manifold is the mount for the carburetor(s); it connects carburetor bores with cylinder-head ports. Ideally, a manifold divides fuel, air, exhaust residuals and recirculated exhaust (EGR) equally among the cylinders at all speeds and loads. Each cylinder should receive the same F/A ratio as all the others for minimum emissions maximum fuel economy and power production.

Manifold passages should have approximately equal length, cross-sectional area and geometric arrangement. For various reasons, this design goal is seldom met. Flow-equalizing features are often used to make unequal-length passages perform as if they *were* nearly equal.

The manifold designer must consider that some liquid fuel is usually moving around on the manifold's walls and floor. Both fuel vapor and liquid have to be distributed so the cylinders' F/A ratio stays the same. Liquid fuel is controlled by sumps, ribs and dams to assist in this distribution.

The manifold floor and carburetor base are typically mounted parallel to the ground to avoid gravity's influence that can cause uneven distribution of the liquid fuel. This usually means angling the carburetor pad/flange because the crankshaft centerline may be angled for drivetrain alignment. The crank is not necessarily parallel with the ground.

Heating the manifold is essential in all but racing applications. Heat helps vaporize liquid fuel. Fuel may not vaporize as it leaves the carburetor. Or, it may drop out of the mixture due

Holley's Pro-Series tunnel-ram manifolds are available for small- and big-block Chevrolets and small-block Chryslers with W-2 heads. This small-block Chevrolet unit is equipped with two Model 4500s, but an interchangeable top is also available for Model 4150s. Small plenum volume and relatively short runners produce maximum horsepower from 4500–8000 rpm.

to an increase in absolute pressure or other reasons.

Small venturis aid mixture velocity through the carburetor and into the manifold. They generally aid vaporization and consequently distribution. High-speed airflow through the carburetor and manifold tends to maintain turbulence, but turbulence *reduces* flow. Velocity helps to keep fuel droplets in suspension, and thereby tends to equalize cylinder-to-cylinder F/A ratios.

If the mixture slows, as happens wherever passage size increases, fuel may separate from the air stream and deposit on manifold surfaces. Var-

iations in the mixture supplied to the cylinders result.

Passage design should give sufficiently fast mixture flow to maintain good low- and mid-range throttle response without reducing volumetric efficiency at high rpm. Passages must be shaped to carry the mixture without forcing the fuel to separate from the air on the way to the cylinder head.

When the mixture is forced to turn or bend, the air turns more quickly than the fuel and the two may separate. When this happens, a cylinder after the bend will receive a leaner F/A ratio than if the fuel had stayed in the mixture stream.

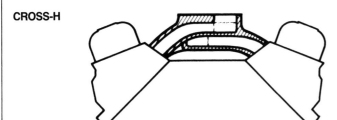

CROSS-H

Cross-H or two-level manifold feeds half of cylinders from one side of carburetor—other half from other side of carburetor. Two sides of manifold are not connected.

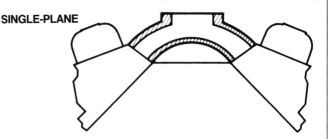

SINGLE-PLANE

Single-plane manifold has all cylinder intake ports connected to a common chamber fed by the carburetor

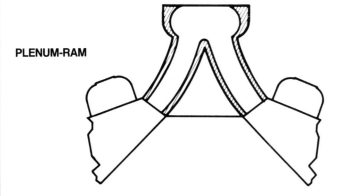

PLENUM-RAM

Plenum-ram manifold has a plenum chamber between passages to intake ports and carburetor/s.

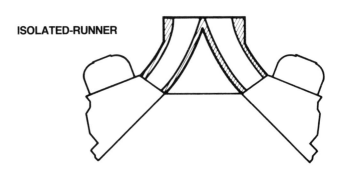

ISOLATED-RUNNER

Isolated-runner (IR) manifold uses an individual throttle bore of a carburetor for each cylinder. There is no interconnection between the intake ports or throttle bores.

Pulsing within the manifold must also be accommodated. This action is called *back-flow* or *reversion*. It occurs twice during a four-cycle sequence. A pulse of residual exhaust gas enters the intake manifold during the valve-overlap period when both the intake and exhaust valves are off their seats. Another pulse reflects toward the carburetor when the intake valve closes.

Either of these pulses may hinder flow in the manifold. In severe cases, pulses pass through the carburetor into the atmosphere. A fuel cloud stands above the carburetor inlet. This is called *standoff*.

Standoff occurs as the fuel/air column starts and stops in the manifold in response to intake valve opening and closing. Pulsing or bouncing movement of the column pushes fuel out of the airhorns. Fuel is drawn back into the carburetor on the next intake stroke. But more bounces back out of the carburetor when the intake

Big-block Chevy 300-4 (oval-ports) or 300-5 (rectangular ports) is a good example of a race single-plane manifold. All cylinders share mixture from plenum cavity immediately under carburetor/fuel-injection mounting flange. Typical of all Holley race manifolds, there's no provision for EGR, choke or exhaust heat. Runners, plenum and carburetor mounting are separated from the tappet-chamber cover by an air gap. This tends to reduce inlet-charge temperature, thereby providing higher volumetric efficiency.

valve closes. The air column pulls fuel out of the discharge nozzle in *both* directions, on its way into the engine, and as it bounces back to the carburetor inlet.

Fuel is metered into the air stream *regardless of the stream's direction*—down through the carburetor or up through it from the manifold. This bi-directional flow occurs to some

Classic Cross-H (dual-plane) manifold design used by Ford engineers for their 427 cid single overhead cam drag racing engine. Vacuum-secondary Holley four barrels were mounted "backwards." Ford Motor Company photos.

degree in most manifolds and in a marked degree in others.

F/A ratio differences between cylinders may be remedied by using unequal jet sizes. Richer (larger) jets are used in a section of the carburetor feeding lean cylinders. And leaner (smaller) jets are used for carburetor barrels feeding rich cylinders. Unequal jetting, commonly called *cross-* or *stagger-jetting,* is a partially effective remedy. In such cases the manifold is at fault.

Stagger-jetting is not very effective with plenum or other single-plane manifolds because all cylinders share the output of all carburetor barrels.

Manifolds also provide mounts for accessories. A manifold can be a tappet-chamber cover and have coolant passages or attachments. Provisions for choke operation can include a mount for a bimetal thermostat. Exhaust-heat passages to the carburetor flange and thermostat mount may be included too. Some have attachment points and holes for distributors. The manifold can also include a mount for an EGR valve and passages to distribute exhaust gas into the manifold.

MANIFOLD TYPES

There are a number of different manifold types:

- Single-plane.
- Dual-plane Cross-H
- High-rise single- or two-plane.
- Individual or isolated-runner (IR).
- Plenum chamber (Tunnel or Plenum-Ram).

Single Plane—Roger Huntington, in a *Car Life* article, "Intake Manifolding" said, "The very simplest possible intake-manifold layout would be a single chamber that feeds to the valve ports on one side and draws from one or more carburetor venturis on the other side. This is called a *common chamber* or *runner (single-plane)* manifold. Common-chamber manifolds have been designed for all types of engines—inline sixes and eights, fours, V8s. It's the easiest and cheapest way to do the job. In fact, most cur-

rent in-line four and six-cylinder engines use this type of manifold."

In the same article, Huntington also explained, "But we run into problems as we increase the number of cylinders. With eight cylinders there is a suction stroke starting every 90° of crankshaft rotation. They overlap. This means that one cylinder will tend to rob fuel/air mixture from the one immediately following it in the firing order, if they are located close to each other on the block. With a conventional V8 engine alternate-firing cylinders are actually adjacent—either 5-7 on the left in AMC, Chrysler, GM firing order, or 5-6 or 7-8 in the two Ford firing orders."

A single-plane manifold may use a divider with a hole in it to connect all of the cylinders. Or, divider height can be lowered to provide a space so all cylinders essentially draw from the same common area.

Smaller carburetors can be used effectively with single-plane manifolds. The common chamber damps out most pulsing, which reduces flow capability. A small carburetor quickens mixture flow and improves throttle response at low- and mid-range speeds.

Dual Runner—(*dual port*) manifolds appear to be dual-plane designs, but they are actually two single-plane

Splendid high-rise, race single-plane manifold example: Holley 300-25 raises carburetor or fuel-injection to give runners straight downward shot to the cylinder-head ports. Air gap under plenum and runners keeps fuel/air mixture cooler for more efficiency. Carburetor is 0-8162 850 cfm double-pumper with secondary idle-mixture screws for improved idle control. Holes in throttles help to compensate for modern racing camshaft profiles. Plumbed with Holley's universal fuel line, Part 34-1.

Removable, interchangeable tops on Holley's 2 x 4 tunnel-ram racing manifold mount two 4500 carburetors (left) or two 4150 carburetors (right). A special top plate is offered for the 2 x 2 Pro-Jection fuel-injection. These manifolds are available for small- and big-block Chevrolets and small-block Chryslers with W-2 heads. This manifold is also shown on page 91.

manifolds. A small- and a large-passage manifold are stacked one above the other in a single casting. The small-passage network is connected to the carburetor's primary side. This supplies high-speed mixture flow for good distribution and throttle response at low- and mid-range rpm. Larger passages are connected to the secondary portion of the carburetor. These supply extra capacity for high-rpm operation.

Experimental dual-runner *two-plane* manifolds were developed by manufacturers including Ford, AMC, International Harvester, Holley Carburetor and the Ethyl Corporation. The primary goal was emission reduction. Volvo's mid-'60s manifold used a similar idea. Mazda rotary (Wankel) engines used the same manifolding concept.

Dual-Plane—(*Cross-H*) manifolds, also called *two-plane,* have been shown to be more throttle- and torque-responsive at low- and mid-range speeds than most single-plane designs. Two single-plane manifolds within a single casting are arranged so each is fed from one half (one side) of a two- or four-barrel carburetor. Each feeds one half of the engine. A Cross-H manifold mounted on V8 is a familiar example.

Each carburetor half is isolated from the other, and from the other half of the manifold, by a plenum divider. Manifold passages are arranged so successive cylinders in the firing order draw first from one plane, then the other. In a 1-8-4-3-6-5-7-2 firing order, cylinders 1, 4, 6, 7 draw from one manifold plane and one half of the carburetor. Cylinders 8, 3, 5, 2 are supplied from the other plane and the other side of the carburetor.

Because there is less air mass to activate during each inlet pulse, throttle response is quicker and mid-range torque is improved—as compared with a conventional single-plane manifold.

Dividing the manifold into two sections causes flow restriction at high rpm. Only one half of carburetor flow capacity and manifold volume is available for any intake stroke. Thus, the divider is sometimes holed, reduced in height, or removed completely. This makes more carburetor and manifold capacity available and increases high-rpm capability.

Holley has several dual-plane manifolds: 300-36, 300-38, 300-48, 300-49 for the small-block Chevy; 300-42 and 300-43 and 300-50 for the big-block Chevy, and the 300-39 for the small-block Ford.

Tuning a Dual Plane Manifold— Bottom-end performance is reduced when the divider is removed or its height is lowered because the volume increase on each side or section decreases mixture speed at low rpm. This is not as severe as with a single-plane manifold, but it is a problem to consider when lowering divider height in a dual-plane manifold.

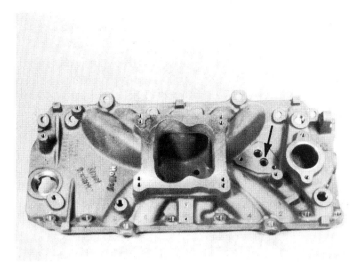

Holley 300-3 street single-plane manifold for oval-port big-block Chevrolet. Universal mounting pad accepts square-flange or spread-bore carburetors without adapter, or Pro-Jection. EGR provision (arrow) in most Holley street manifolds accepts bolt-on or clamp-on EGR valves. Plugs seal passages when no EGR valve is required. Plugs drive in easily and a little sealant helps to prevent leaks. Overall height was kept low to clear modern hood lines. Photo at right shows clamp-on type EGR valve.

On engines built for high-rpm operation, divider removal can be a tuning plus. Street and track applications combine careful selection of camshaft and other engine pieces to build torque into an engine. Leave the divider in place. The super-tuner can remove small amounts of it to reach the optimum height for a particular engine/parts combination and application. For street performance where 4000 rpm is rarely exceeded, a well-designed dual-plane manifold is hard to beat.

High-Rise or High-Riser?–Optional *high-rise* high-performance manifolds have been offered by some automobile manufacturers. Several aftermarket manifold makers have also made them. A high-rise manifold aligns the cylinder-head port angle with that of the manifold passage or runner. A nearly straight, downward path connects the area under the carburetor to the intake port. In a V6 or V8 engine, the entire network of manifold runners is raised, along with the carburetor mounting.

High-rise should not be confused with *high-riser,* which refers to a spacer between the carburetor base and the manifold. It makes a longer riser and straightens flow disturbed by the throttle plate—before it enters the manifold. The spacer improves distribution by eliminating directional effects caused by partially-opened throttles. However, unless the spacer is divided to keep the two planes of the manifold separated, the spacer creates a single-plane manifold.

High-rise and high-riser designs may be used with either single- or two-plane manifolds.

Independent Runner (IR)–Independent- or isolated-runner manifolds are for racing applications. One carburetor throat and one manifold runner serve each cylinder-head inlet port. Each runner and throat combination is isolated from its neighboring cylinders.

Carburetors used with IR manifolds have fuel-metering for each barrel because cylinders cannot share functions (except accelerator pumps). There's a separate idle and main system for each barrel.

Sharp pulses occurring in two directions in an IR intake system start main metering operation very quickly. A large accelerator pump isn't required to cover up nozzle "lag." A power valve can't be used. It won't function correctly under the extreme pulsing conditions.

Inside plenum of 300-7 street single-plane manifold for small-block Chrysler (stock heads) or 300-21 (W-2 heads). Vanes help cylinder-to-cylinder distribution.

An IR manifold allows tuning to take advantage of *ram effect.* This is an important point in manifold design. The shorter the tuned length, the higher the rpm at which peak torque occurs.

Carburetors used with IR manifolds must be *much larger* than for other manifold types. This is because (1) each cylinder is fed by only one carburetor throat or barrel and (2) the pulsing reduces carburetor flow capacity in the rpm range where standoff becomes severe.

IR manifold/carburetor combinations require standoff containment.

Looking down throat of deep-breathing race single-plane manifold. Note straight path for inlet charge from carburetor or fuel injection to cylinder head ports.

Holley 300-22 manifold for 1975 and later Buick, Pontiac and Oldsmobile V6s and 1966-77 AMC Jeeps. This is called a *street single-plane,* even though the divider comes almost to the bottom of the carburetor.

Chrysler low-block 361–400-cid engines get a performance boost with the 300-10 street single-plane manifold. Works with either square-flange or spread-bore carburetor or Pro-Jection.

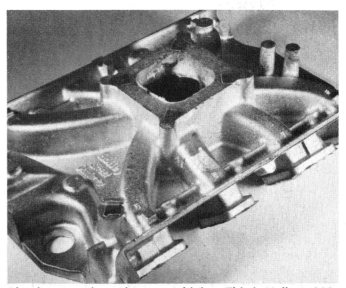

Aluminum casting prior to machining. This is Holley's 300-11 Street Contender manifold for the Ford big-block 352, 390, 428 engines with or without EGR. You'll need a crane to get the original equipment manifold off engine because it weighs about 100 pounds.

Usually, a stack long enough to hold standoff is placed atop the carburetor inlet.

Some tuners say the IR carburetor venturi should be at least as large as the intake-valve diameter. Holley made two Model 4500 carburetors specifically for IR use: List 6214 or 0-6214 (no longer made) and 0-9377. Even with their rating of 1150 cfm,

two of these don't have enough capacity for a 302 cid V8. Holley has calculated that this engine would require two 1650 cfm four-barrels for IR.

Plenum Ram–This version of a single-plane manifold has a plenum chamber between the carburetor base and manifold runners. The chamber helps dissipate the strong pulsing, so less pulsing enters the carburetor to

disrupt flow. Just as important, it allows the cylinders to share carburetor flow capacity. In the typical dual-quad plenum manifold, three or four cylinders simultaneously draw mixture from a plenum being fed by eight carburetor bores.

Sharing carburetor capacity makes the plenum-ram "forgiving" of carburetor capacity. This is true so long as

Holley's manifold installation kits are complete for simple and trouble-free installation. You may find parts not needed for your particular engine. This one is for Chevrolet small-block V8.

Holley 300-7 street single plane manifold for Chrysler 273/318/340/360-cid engines. Flange accepts both spread-bore and square-flange carburetors or Pro-Jection throttle body.

the carburetors are not *too large* for the desired rpm range.

The typical Pro Stock drag racer runs at 6800–8500 rpm. With a large-displacement engine, there is usually enough airflow at 6000 rpm to start the main systems, even with a pair of the largest 4500s.

The smaller the plenum cavity, the smaller the main-jet requirement. A smaller cavity sharpens pulsing to the carburetors. The main system starts sooner and more fuel is pulled out of the jets because of the pulsing.

MULTIPLE CARBURETION

Until about 1967, two or more carburetors were considered essential for any modified or high-performance engine. No self-respecting auto enthusiast would consider building an engine with fewer than two carburetors–unless the rules required it. That has all changed. One four-barrel is usual and accepted for *all* street and many competition applications.

Sophisticated two- and four-barrel carburetors with small primaries and large progressively operated secondaries, are matched with single-carburetor manifolds for excellent performance and low emissions.

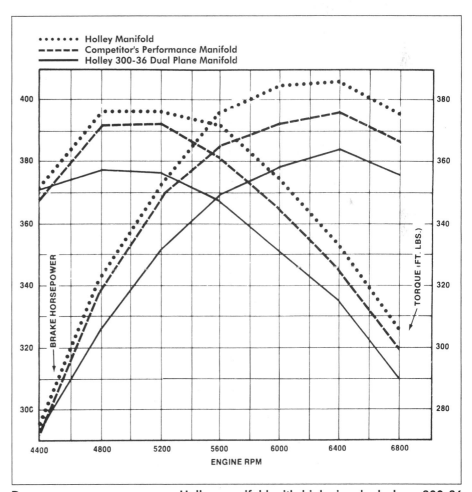

Dyno-power curves compare Holley manifold with high-rise dual-plane 300-36 Holley manifold and competitive performance manifold. Engine is 350 cid Chevrolet LT-1 with performance headers and 0-4781 (850 cfm double-pumper) carburetor used with "best power" jetting for each manifold.

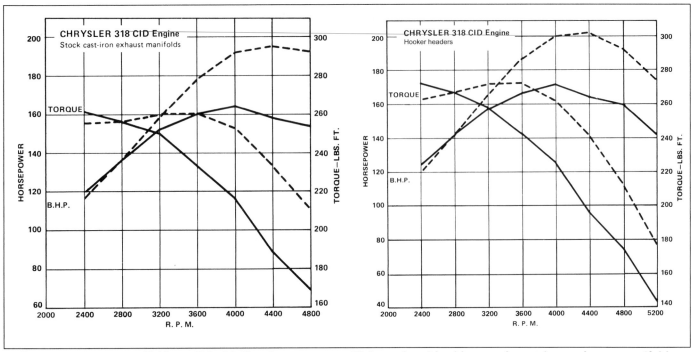

Holley 300-7 Street manifold on 318-cid Chrysler gave over 30 hp gain with either stock cast-iron exhaust manifolds or headers. Headers kept power up at higher rpm. Solid line in both charts is original cast-iron intake manifold, 2-bbl. carb with no air cleaner. Dashed line in left chart represents 300-7 manifold with Model 4160, 0-7009 carb, 66–73 jets, and open-element air cleaner. Dashed line in right chart is same combination, but using 64–71 jets.

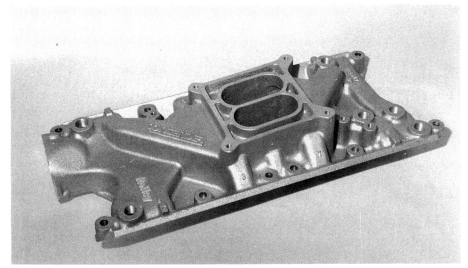

Low-cost, low-rise dual-plane Ford manifold is Holley 300-39. Fits square-flange carburetors or Pro-Jection. For small-block Fords 221–302 cid without EGR. Provides 21 hp more than stock at 3600 rpm.

An application remains for multiple carburetion, at least as we once knew it: drag cars and boats. It is also useful for extended high-rpm operation at California dry-lakes events and the Bonneville Nationals Speed Trials.

Multiple carburetion is used to reduce the pressure drop across the carburetors to a minimum. This ensures the least possible hp loss from this restriction. Of course, low-rpm operation in these applications isn't a consideration. The extra venturi area and carburetor flow capacity don't create any problems.

A great many '60s and '70s engines were factory-equipped with two- and three-carburetor manifolds. By 1973 all U.S. manufacturers equipped their high-performance engines with single four-barrel carburetors. The majority of these single- and multi-carb factory manifolds used Holley two- or four-barrel carburetors with diaphragm-operated secondary throttles–or secondary carburetors. Holley has reissued many of these original part numbers for Chevrolet, Chrysler and Ford muscle-car restorations. See your Holley dealer or the Holley parts catalog for specifics.

If you're the proud owner of a car with 2 X 4s or 3 X 2s, avoid buying mechanical linkages to open the throttles simultaneously. Leave the diaphragm and progressive operation intact on the end carburetors.

The need for adequate mixture velocity is stressed throughout this book. We explain how to obtain it by selecting the correct carburetor capacity for engine size and rpm.

Simultaneous opening of multiple throttles goes against these recommendations. The car is hard to drive because velocity through the carburetors is drastically reduced at low and medium speeds. Progressively operated secondaries are more desirable and improve driveability.

AMC 304–401-cid V8s can use 300-31 Z-series street manifold. Because full plenum divider contains an opening, this is not a *true* dual-plane manifold. Shown with 600 cfm Model 4160 0-8005 that has vacuum secondaries, two-stage power valve, Ford kickdown linkage and electric choke. This is an emission-type carburetor.

HOLLEY INTAKE MANIFOLDS

Back in 1976 Holley introduced a new and complete line of intake manifolds. These manifolds resulted from intensive engineering and development. Dynamometer, emissions, fuel economy, street driving and drag-strip testing were all used to create these manifolds.

Runners were designed for minimum flow loss using cross-section profile analysis. Infrared exhaust-gas-analysis instrumentation accurately measured cylinder-to-cylinder fuel distribution. Good distribution was one of the paramount considerations in making manifolds that would not alter original-equipment emissions performance.

Holley considers the carburetor and manifold or fuel injection and manifold as a combination. Holley's carburetors and fuel injections include design features to optimize performance of the total induction system.

Holley's manifolds have been named *Contender, Street Contender, Strip Contender, Z-Series, Dominator, Pro Dominator, Pro Series* and *Holley.* Holley's 1994 Performance Catalog listed essentially the same group of manifolds as *Street-Legal Dual-Plane, Street Performance Single Plane, Competition (Single-Plane), and Pro-Series Tunnel Ram, Dual-Plane, Street—Single-Plane, Race—Single-Plane and Tunnel-Ram.* Avoid being confused by names or terms by consulting the current catalog.

On the Street—Holley engineers began with a "clean slate" to design a line of street manifolds. Some Holley street manifolds have universal carburetor flanges that mount either square-pattern or spread-bore carburetors without an adapter. Choke-operating features are also included.

Many of the street manifolds have full emission control provisions, including EGR. Tapped holes are provided for operating power brakes and other accessories. Exhaust-heat passages ensure fast warmup and good mixture distribution. Flange heights are kept as low as possible to fit under low hood lines.

The street-manifold development effort emphasized making the manifold/fuel-injection/carburetor supply good power with stock cast-iron exhaust manifolds. Further testing ensured that the manifold/fuel-injection/carburetor teams would also work well with headers. Some of Holley's street manifolds work with both spread-bore and square-flange carburetors. Incidentally, all Holley manifolds are 100% pressure-tested.

The installation kits are the most complete ever offered. Installation instructions are written so inexperienced mechanics can install a manifold and make it work correctly.

See the graph on page 97 for performance comparisons between the Holley and another manifold. Each used an 850 cfm 0-4781 double-pumper carburetor. The Holley showed a 12-hp advantage on a 350 cid Chevrolet engine.

On the Strip—Holley's Competition Manifolds are designed for speeds above 4500 rpm. An air gap between the manifold valley cover and the intake runners reduces intake charge heating from engine heat. Better power results from the cooler and denser intake charge to the cylinders.

These are true *racing* manifolds. Runner configurations and plenum volumes are optimized to favor all-out performance. Development and testing included dynamometer and drag strip work with actual competition engines. Development was conducted with headers to optimize power production above 4800 rpm.

These manifolds are not designed for street use. There is no provision for chokes, EGR or other emission-control equipment.

Pro-Series Tunnel Ram—Three of these manifolds provide coverage for Chevrolet small block and Chevrolet big block, plus the Chrysler small-block with W-2 heads. Interchangeable tops are available: one fits 4150 carburetors and the other is set up to work with the 4500 carburetors. These manifolds are specifically suited for the 4500—8500 rpm power band.

HOLLEY DUAL-PLANE

Manifold experts and would-be experts have always argued the merits of single-plane versus dual-plane manifolds. Advocates on either side can make test data work in their favor. It is safe to say that single-plane man-

Oval-port Chevrolet big-block engines can use this high-rise dual-plane manifold 300-43 equipped with a 750 cfm Model 4500 Dominator carburetor. 300-42 fits the same heads and accepts square-flange carburetors or Pro-Jection. These manifolds work from idle to 6000 rpm. They include a divorced-choke provision, a heat crossover for cold-weather driveability and most original equipment hookups.

Holley 300-36 high-rise dual-plane manifold for the small-block Chevrolet copies Chevrolet's famous high-performance high-rise manifold used on the 370-hp 350-cid LT-1 engine in the late '60s. This superb street-performance manifold has good low-rpm torque, yet winds out with great performance to 7000 rpm. Use with a 4150, 4160 or 4010 carburetor or Pro-Jection. See dyno charts on pages 97, 101.

ifolds tend to favor rpm and horse-power. Dual-plane types serve best at low- and mid-range rpm and are basically *torque* manifolds.

To make their manifold line complete, Holley added dual-plane manifolds in the early '80s. The first ones were for the Ford and Chevrolet V8s, followed by V6 manifolds to fit the Buick, Chevrolet, Oldsmobile and Pontiac.

Perhaps the first manifold to consider is the 300-48 Street Legal manifold for the 262—400 cid Chevrolet small blocks. This high-rise dual-plane manifold fits all years except 1987 and later Chevrolet cast-iron cylinder heads with the canted center bolt. It includes EGR provisions and late-model alternator and A/C bracket provisions. This manifold is unusual in that it is emission-legal in all 50 states of the U.S.

The 300-36 is a high-rise dual-plane manifold very similar to the original Chevrolet LT-1 manifold. Match it with a square-flange four-barrel and an open-element air cleaner and you've got a superior street package with good driveability, fuel economy and performance.

There is no EGR provision, so emission regulations must be considered before installing it. The manifold is about 1-1/4-in. higher than the stock cast-iron manifold, so measure hood clearance before you buy. A Pro-Jection unit with a lower profile may fit when a carburetor won't.

The 300-38 is a shorter dual-plane manifold for the small-block Chevrolet. Its low profile means there's no hood-clearance problem. Because of the smaller, flatter-angle runners, both power and torque are reduced as compared with the 300-6. There is no EGR provision. Use a 600-cfm square-flange or spread-bore four-barrel or a two-barrel Pro-Jection and an open-element air cleaner. This is the lowest-cost Holley V8 manifold for the Chevrolet.

For Ford small-blocks, the square-flange 300-39 follows the same design approach as that used for the 300-38. It's also a very good low-cost dual-plane manifold with no hood-clearance

problems. It fits the 221, 260, 289 and 302-cid engines (except Boss).

Two Holley dual-plane street-legal manifolds are designed specifically for Pro-Jection fuel injection on the 262—400-cid Chevrolet small blocks. These two manifolds will fit all years except 1987 and later Chevrolet cast-iron cylinder heads with the canted center bolt. The 300-49 is designed specifically for the two-barrel Model 3200 throttle-body injector flange. The 300-50 is designed specifically for the four-barrel Model 3400 throttle-body injector flange.

In the mid-'80s, Holley Engineering used Zora Arkus-Duntov of Corvette fame as a consultant to develop a line of "Z" manifolds. These use a balance-tube passage connecting the two rear ports to minimize the problem of the two rear cylinders on the left side (5 and 7 in GM engines) robbing each other of intake charge because they follow in the firing order. The left and right sides of the plenum chamber are separated by a divider—so you can call them *dual-plane* manifolds. The

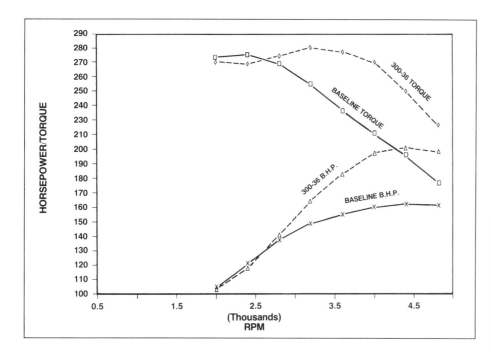

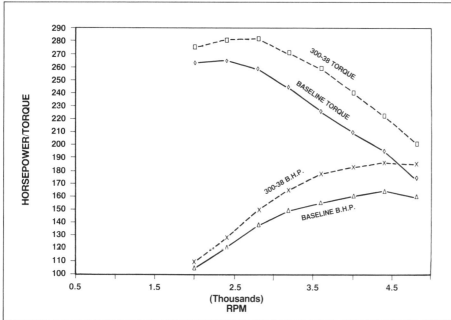

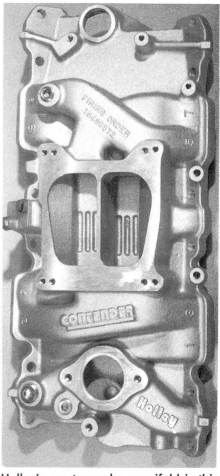

Holley's most-popular manifold is this 300-38 dual-plane for 1957 through present small-block Chevrolets without EGR from 265–400 cid. Universal mounting flange accepts square-flange and spread-bore carburetors and Pro-Jection. Low profile reduces hood-clearance problems. Includes divorced-choke pad and provision for alternator and air-conditioner bracket mounting.

Dyno comparisons. At top are HP and torque measured on a 1983 305 cid Chevy V8 with no air cleaner, stock intake manifold, Rochester Q-jet and Stahl 1.75-in. headers (solid line) vs. same engine with no air cleaner, 300-36 dual-plane high-rise manifold, 0-1850 carburetor and same Stahl headers (dashed lines). At bottom is same stock engine configuration compared with low-rise dual-plane 300-38 manifold, 0-1850 carburetor and same Stahl headers (dashed line).

result is a compromise between a single- and dual-plane manifold. Low-rpm torque is increased over a single-plane and high-rpm power is increased over the dual-plane. These

manifolds are now made only as the 300-31 for AMC 304–401-cid V8s and the 300-29 for Chrysler 273–360-cid V8s. But if you should happen to find one for a small-block

Chevrolet, Chrysler or Ford at a swap meet and it fits your engine, it's an especially good torque manifold.

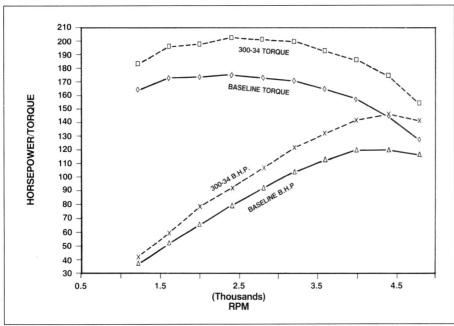

Dyno comparisons on 1980 229 cid Chevy or Pontiac V6. Solid lines are hp and torque of stock engine with stock intake manifold and Rochester 2-bbl. carb. Dashed lines are same engine with 300-34 street single-plane manifold with 450 cfm four-barrel. No air cleaner was used for either of these tests.

Chevrolet 90-degree V6, either 200 or 229 cid from 1976–80 can be equipped with Holley's 300-34 mani-fold. Recommended carburetor is the 450 cfm Model 4360 0-9694, which is specifically calibrated for this engine/manifold combination. Air cleaner is Holley's Part 120-103 gold-finished 9-in. diameter air cleaner.

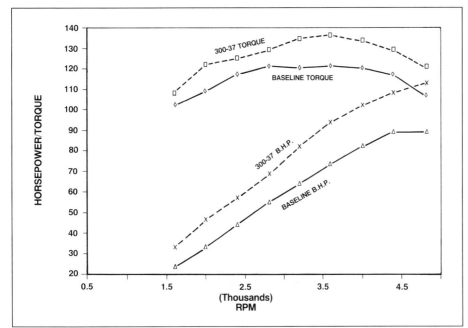

Holley dyno comparisons on 1979 2.2-liter 20R Toyota inline 4 cylinder. Solid lines are HP and torque of stock engine with stock air cleaner, intake manifold and Aisan 2-bbl. carb. Dashed lines are outputs of same engine with stock air cleaner, Holley 300-37 intake manifold and Model 4360 0-9973 carb.

Toyota 1977-80 2189cc engines can use this 300-37 intake manifold. Model 4360 carburetor (not described in this book) shown here is specially calibrated for this application. Manifold also accepts a square-flange Model 4150/60 carburetor for highly modified engines. Using this manifold requires extensive throttle-linkage modifications.

How to Install a Manifold 6

Photos by Alex Walordy

Now that we've led you to the right manifold and carburetor or fuel-injection combination, we'll help guide you in the installation.

If you are an experienced racer or enthusiast you've probably changed many manifolds in your time. This section is not for you. We've included it for the novice who might be doing it for the first time. The following was done with the help of good friend Alex Walordy who has been writing magazine articles and books about Holley equipment for more years than either of us will admit to. We appreciate his expertise on the step-by-step photos. Thanks, Alex!

THE INSTALLATION PROCESS

It's easy to change manifolds. Installing the extra Holley power can take anywhere from 45 minutes to several hours, depending on the amount of accessories and the number of visitors and cups of coffee.

Jim Patrick from Mayfair's Automotive in Taylor, Michigan, helped us go through all of the tricks needed to speed up the job and avoid mistakes. We used one of his 350 cid small-block Chevrolets as the example.

We made this installation on his dynamometer because it was so much easier to photograph critical details. Too many things get in the way of the lens in an actual vehicle installation.

Getting Ready—Very little is involved in the way of tools. If you are careful, there won't be any fuel, air or oil leaks. A manifold can be changed

1. Intake-manifold installation is fairly simple if you know ahead of time what to expect. Follow these tips to do it right the first time.

right in the vehicle. You don't have to remove the hood, although it's faster and easier if you do.

Throw on a pair of fender covers and go to work. To avoid any shorts or arc-overs, disconnect the battery

cables. Now clean up the top of the engine, both for convenience and to prevent dirt from getting in the engine when the manifold is removed. Coolant must be drained because you'll be disconnecting hoses and thermostat. To avoid a messy floor, clean the engine and drain its coolant outside the shop.

Label the Connections—On later-model vehicles you are faced with a large number of vacuum hoses leading to emission controls. Rather than trace them all down, get a roll of masking tape. At each hose you remove and at each connection it came from, stick on a flag of doubled-over tape, marked with identical numbers. During assembly, just match the numbers.

Air Cleaner—When you remove the air cleaner, find the vacuum hoses that lead to it on the underside, and tag them also. Now is a good time to clean the air cleaner. Check the breather element. Hold it up to the light and replace it if it's dirty. Replace the gasket that goes between the air cleaner and the carburetor or fuel-injection unit. That gasket at the top of the unit keeps the engine from getting sanded by dirty air that is bypassing the air cleaner.

Throttle Linkage—This step alone is sure to eliminate wasted hours of tracking down mystery mistakes. Take the time to trace the throttle linkage. It may include automatic-transmission and cruise-control connections. Make a sketch of these connections. And sketch the location of different brackets retained by the manifold bolts. The mechanic at a dealership knows all this by heart and your buddy who offers to rip apart the top end will tell you he knows *exactly* where everything goes. It is best to make the sketch. Come reassembly time your buddy may be away on a week's vacation—or maybe he's taken a new job in Hawaii. It's easier to take notes and make sketches or Polaroid photos to record how things looked before you take it all apart.

Removing the Distributor—Before you remove the distributor ignition wires, clean them and tag them. For a few dollars you can buy a plastic gadget with a T handle to help pull the spark-plug boots without ripping them. Yanking on the wires can

2. Remove the distributor connections so you'll be able to take the cap off.

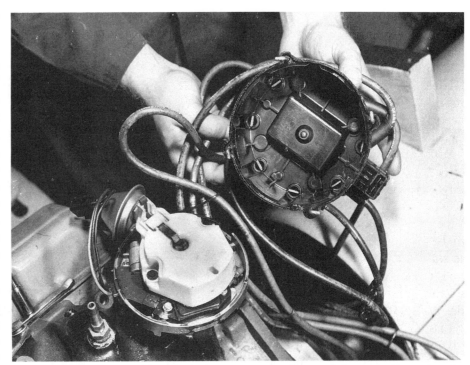

3. Remove the cap and ignition wires to get them out of the way. Note that the cap has a locating tang, and two clips that must be pressed down with a screwdriver and turned. Record and mark the position of the distributor rotor so it can be reinstalled the same way, or bring the rotor to the #1 firing position by pointing it at the #1 terminal on the cap. Scribe the position of the distributor base in relation to the engine block.

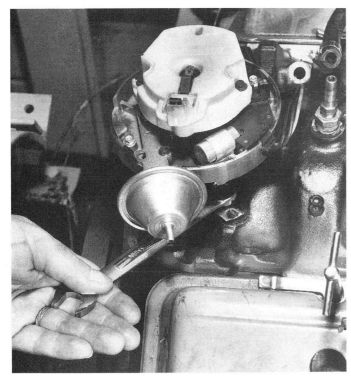

4. Now loosen the bolt at the distributor clamp.

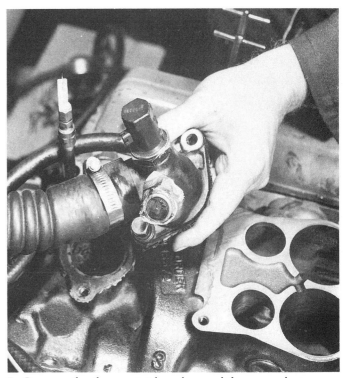

5. Remove the thermostat housing and the upper hose.

destroy a good harness. If you clean and number the wires you won't have to listen to your engine popping and banging when they are inadvertently crossed. If the insulation is hard and cracked, replace the whole set of ignition wires.

Before you remove the distributor, scribe two marks on the distributor body: (1) to show where the rotor is pointing, and (2) to show the position of the distributor in the engine. Those will be a big help so you can reinstall it the way it came out. Another alternative is to turn the engine until the rotor is pointing at the number-one plug wire and the damper (crankshaft pulley) is lined up on the timing marks. Anything that simplifies reinstalling the distributor is good. The housing is retained by one bolt and clamp. A helical gear drive on the distributor causes the rotor to rotate slightly as you pull it out. Note which way it moves as it comes out, and begin with the rotor in that position to reinstall it.

Removing the Carburetor—The carburetor is the next item to come off. This means disconnecting the fuel line,

6. Remove the manifold, taking care to avoid any dirt or debris falling into the engine.

(always use two wrenches on a connection—preferably tubing wrenches). Remove all vacuum lines and the wires to the choke, solenoid or throttle-position switch. The upper radiator hose and thermostat housing are next, fol-

lowed by ground wires, spring brackets and accessories.

Removing the Manifold—Pulling the rocker covers helps to gain extra swing room for the manifold. Now is time for a little extra clean up around

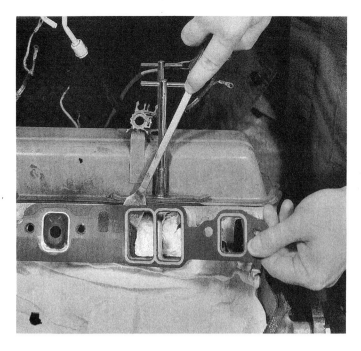

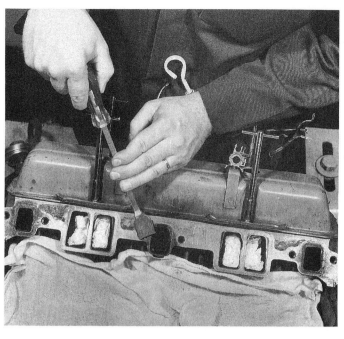

7. Cover the tappet valley with rags and plug the ports with wadded paper to prevent dirt entry. Then use a scraper to peel away the old intake gasket. Use your shop vacuum to suck up the dirt.

8. Carefully remove any carbon traces around the parts nd exhaust-gas crossover. A sharp scraper will speed up the work. Be careful not to scratch the surfaces. The mating gasket surfaces on the block must be scraped clean and then wiped with a thinner-soaked rag.

the manifold so you don't introduce any extraneous dirt into the engine. Use your shop vacuum to get out any dirt around and on the manifold.

After removing all of the intake-manifold bolts, pry out the manifold itself. Gaskets and sealers form a pretty good bond. Just make sure all of the bolts are out before you begin prying because you could break the manifold if one or more bolts are still in place. Take care not to damage any of the sealing surfaces.

More Clean Up—Once the manifold is off, give the cylinder heads and tappet valley a good wash job. Remember to change the oil and filter before starting the engine. Installation success depends on how well you clean up the gasket areas and the rails at the front and rear of the block. Your best bet is a good nick-free scraper. When cleaning gasket surfaces, cover all of the tappet valley with shop towels to protect the engine. Use shop air to blow out all screw holes and the intake-port area.

Preinstallation—Preinstall the manifold dry without end-rail gaskets. Let it rest on the intake gaskets at the heads. Use light and feeler gauges to check how much clearance is available between the manifold and the rails at the front and rear of the block. Normally everything is fine, but if the sides of the heads or the manifold are out too far, the manifold will sit low and there won't be enough room for the end rail gaskets. Then the manifold just won't seal. Now is the time to determine whether a little machining is needed at the rails. Another solution is to use double intake gaskets. Clearance from 1/8 to 1/4 inch at the rails is OK. Now you are ready for final assembly.

Putting on the Manifold—Apply a light coat of cement to the cylinder heads and position the intake-manifold gaskets. They should stick in place and be centered as closely as possible. Next, apply a light bead of silicone sealer around each port opening in the gasket and at the coolant crossover passages. It is good practice

9. Lay a set of intake gaskets in place on the heads with no rail gaskets. Now use a feeler gauge stack to see how much clearance there is at the end rails. 1/8 to 1/4 in. is OK. When there is too little clearance at the end of the manifold to accommodate the end-rail gaskets, milling this area of the manifold may be necessary. With too much clearance the sides of the manifold must be milled. Check the alignment of the intake manifold holes with the threaded holes in the heads.

10. The best rail gaskets come pre-coated with adhesive. Remove the tape and apply the gasket.

11. Position the rail gasket. Depending on how the manifold fit was measured in step 9, in some installations there will only be room enough for a bead of silicone and no gasket. Remove the tape and apply the gasket.

12. The easiest way to apply sealer to intake-manifold gaskets is to spray on an even coat. This tacky material will help the gasket stick to the head.

13. Cement the intake manifold gaskets in place on the heads. You don't want them moving when the manifold is installed. The best gaskets have their own printed sealing bead. Use some silicone sealer applied in a light coat at critical areas. Large lumps of sealant can prevent the manifold from sealing.

to center-punch little stake marks along the rails to give the silicone a better grip. If you aren't using rail gaskets, apply a thick enough bead to the end rails so it will squash down and straddle the rail when the manifold is tightened. This positions the bead and will not let it escape. Apply one continuous bead to the side rails from head gasket to head gasket. The other alternative is to use rail gaskets along with a little silicone sealer.

When the manifold is installed, bolted down and tightened, use a mirror to check the rear rail for leaks or gaps. Intake-manifold bolts should be lightly snugged down from the center out. Retighten them all evenly a couple of times in increasing torque steps.

Installing the Distributor—Now we're ready to reinstall the distributor.

14. When you install the gasket, align it to the bolt holes. Then check for port alignment. For racing engines the gasket can be used as a template to match the cylinder head and manifold ports.

15. A small glob of silicone sealer is used at the corners between the rail gaskets and the tangs of the manifold gasket. You may want to coat the top of the rail gaskets with a thin bead of silicone.

16. Carefully set the manifold into place so the gaskets are not disturbed.

17. Torque the intake manifold bolts in several steps. There may be some spots where a torque wrench cannot be used.

A new paper gasket should be used at the base. When you removed the distributor, the helix or spiral lead on the distributor gear caused the rotor to turn. Now, when you install the distributor, the rotor has to be displaced by a matching amount so it can spiral back (in the opposite direction) to the correct position. If you scribed the marks at the rotor, distributor base and block, and they again line up, just clamp the distributor down. This should allow you to start the engine. Use a timing light to set timing accurately.

Installing Carburetor or Fuel Injection—Before you install your new Holley carburetor or fuel-injection unit, use a straight edge to check the top of the manifold flange. If it is warped, a little filing may be in order. We recommend gasket 108-12

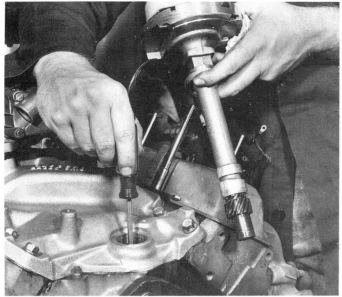

18. The oil-pump shaft may have to be turned with a screwdriver to let the distributor slide all the way down.

19. Install the distributor and tighten down in the original position. If the engine pops on start up, check the firing order versus the spark plug wiring. You may have crossed the wires. Use a timing light for the final ignition-timing setting.

20. Tighten all four corners of your Holley carburetor or throttle-body injection evenly and in a criss-cross fashion. Avoid stacks of soft gaskets.

21. Set the fuel level, connect the throttle linkage and set the idle.

for square-flange applications such as the Holley models 4150, 4160 and 4010. For spread-bore versions such as models 4165, 4175 and 4011 use gasket 108-25. These gaskets are thicker with phenolic inserts at the corners. The added thickness offers heat insulation. Also, you should make sure that the gasket seals against the manifold flange just in case there are a few areas under the carburetor unsupported by the manifold flange. Because of this slightly added thickness, check hood clearance. In the case of fuel injection, use the gasket provided in the kit or an exact replacement and follow the instructions included in the kit.

The throttle-body attachment nuts should be tightened lightly and then snugged down in a criss-cross pattern. Do not overtighten lest you break an ear on the throttle body or cause binding of the secondary shafts on diaphragm-operated carburetor models. We do not recommend a stack of soft gaskets without the corner inserts.

Check the Linkage—Throttle linkage and throttle cables should work smoothly, without sticking, and should allow the throttle to open all the way, or to close fully without the need for a super-stiff spring. With the engine off, have someone in the car operate the accelerator pedal. Check the idle and wide-open-throttle positions and adjust the cable or linkage if necessary.

At wide-open throttle, manually operate diaphragm secondaries to make sure the throttle shaft is not binding.

The last move is to reconnect the choke, fuel line, plus all of the accessories and hoses. If you are using a Holley electric choke controlled by the ignition switch, you will need a 12-volt supply. One good source is the accessory terminal of the fuse box. Again, with fuel injection, refer to the installation manual.

Fuel Line—The original 5/16-in. fuel line is borderline for a high-output combination, especially at high altitudes. You may want to change to a 3/8-in. line, at least from the fuel pump to the carburetor. Now is the time to install an inline filter if there isn't one already. We recommend the see-through type. This is good insurance against flooding due to foreign material lodging between the needle and seat.

Any fuel line should have good support. Route lines to avoid all hot areas and do not use right-angle fittings which cause restriction. These actions will help prevent vapor lock.

TUNING

Any new fuel system and manifold installation calls for a little tuning. You can't just remove manifold A and install manifold B and go run a direct comparison. Check the plug readings and the feel of the car. With carburetors, you may need a little more pump shot and a change of jetting by one or two sizes. You may also want to test with and without a spacer under the carburetor, assuming hood clearance permits this. Also, try different ignition timing settings. With injection, follow the tuning method in the installation booklet. A fuel injection or carburetor and manifold work as a team and need to match each other.

DETECTING MANIFOLD LEAKS

An instant indicator of a manifold leak is a rougher-than-normal idle with the throttle at the factory-set curb-idle position. You may hear a hissing or whistling, but this can be masked by other engine noise. The roughness at idle is caused by the engine receiving more air than it needs, creating a lean mixture.

Air leak symptoms appear more at idle than higher engine speeds because the engine is consuming so little fuel/air mixture. Additional air drastically alters the fuel/air mixture ratio. At higher engine speeds the air leak is small compared with the total fuel/air mixture. Consquently, the leak seems to disappear.

So how do you find the leak? First, inspect all hoses attached to the carburetor base or to the intake manifold. Make sure no hoses are cracked or broken. Take special care to check the underside or hidden portions of the hose where a leak might not be obvious. Make sure all manifold-vacuum ports have plugs installed.

Some mechanics spray a little solvent (carburetor cleaner or starting fluid) on hoses or the manifold to locate leaks when the engine is idling. The combustible solvent enters through a leak and the idle smooths and increases about 200 rpm. Use this technique to check for worn throttle-shaft bores. Shafts and bores wear with use. An air leak and lumpy idle result. Spray the solvent directly at the throttle shaft where it enters the throttle body and note the idle response.

Use extreme care if applying this detection method because of the fire hazard of spraying combustible droplets on a running engine.

Other mechanics locate air leaks with propane. The gas can enter a small leak more easily than a liquid, and the fire hazard is reduced because the gas rises and leaves the work area.

Pinpoint any leaks by pointing the propane torch tip at the suspected area on an idling engine. Open the gas valve on the torch. An air leak will draw the gas into the manifold. Note the idle response as specified above.

Neither of these detection methods is a foolproof diagnostic procedure. Solvent or gas vapor can be sucked into the air cleaner or carburetor air horn and the idle will increase from this. You'll be misled into thinking you've found an air leak.

Engine Variables 7

Testing programs require a dynamometer to measure effects of one engine variable at a time. This Chevrolet V8 engine is installed on Superflow SF-901 Computerized Engine Dynamometer System. Air inlet atop Holley carburetor measures airflow. Fuel flow is also measured so printout contains brake specific fuel consumption data, as well as horsepower and torque.

Many variables affect the operation of the carburetor and engine. Their relationships are examined here so you can see how each variable affects the others. Much of this information is important in understanding and tuning stock and racing engines.

SPARK TIMING

Late model cars have engine compartment labels specifying engine idle speed and spark advance settings. Carefully worked out by the factory engineers, these ensure that the engine in that vehicle will have emissions within specified limits.

Controlling Advance—In the early '70s, the trend was to detune the engine to meet emission requirements. Part of the detuning included retarding the spark at idle. Some emission control systems locked out vacuum advance in the intermediate gears. A temperature-sensitive valve allowed spark advance if the engine started to overheat and during cold operation. The advent of the catalytic converter in 1975 allowed more spark advance.

Three-way catalysts, discussed on page 143, limit HC and NOx in the exhaust stream, not by altering engine operation. This allows using more

111

spark advance at idle and part throttle. Driveability and fuel economy are improved because engine efficiency is improved.

With the increased use of electronics to control engine operation, spark advance is even more closely controlled. The ECU receives inputs on engine speed, load, temperature and throttle angle and so on. It accurately changes spark advance to satisfy engine requirements based on the various sensor inputs.

Most carburetors have ports for timed spark and for straight manifold vacuum. Some Ford applications used a combination of venturi vacuum and manifold vacuum, called a *pressure-spark system*. Use of this ended in about 1972.

In general, a retarded spark reduces NO_x by keeping peak combustion pressures and temperatures at lower values than those generated by an advanced spark setting. It also reduces HC emissions. A retarded spark is good for reducing emissions, but bad for economy, driveability and coolant heating.

Fuel is being burned in the engine, but some heat energy is wasted as it flows into the cylinder walls. Because fuel is still burning as it passes the exhaust valve, the exhaust manifolds have to cope with more heat. This aggravates heating problems in the engine compartment. The cooling system has to work harder. Fuel is still burning as it passes the exhaust valve, so the engine's thermal efficiency is less because energy is wasted.

Using retarded spark requires richer jetting in the idle and main systems to get decent off-idle performance and driveability. A tightrope is being walked here. The mixture must not be allowed to go lean or higher combustion temperatures and NO_x will be produced. If mixtures are richened too far in the search for driveability, CO emissions increase.

Because retarding the spark hurts efficiency, the throttle plate must be opened farther at idle to get enough mixture in to keep the engine running. This must be considered by the carburetor designer in positioning the idle-transfer slot. High temperatures at idle and high idle-speed settings also promote dieseling.

Dieseling–Also called *run on*, is primarily caused by greater throttle opening. It is aggravated by higher average temperatures in the combustion chambers. Higher temperatures tend to cause any deposits to glow so self-ignition occurs.

Consider a car equipped with an *anti-dieseling solenoid*. Turning the ignition ON energizes the solenoid so the throttle is moved to its idle setting. Turning the ignition OFF de-energizes the solenoid to change the idle setting to about 50 rpm slower than a normal idle setting. This throttle idle position, that is more closed than normal idle, reduces the tendency to run on. If the engine can't ingest enough mixture to continue running, dieseling won't usually occur.

Timed Spark Advance–Carburetors equipped with timed spark advance (no advance at closed throttle) have a port in the throttle bore. This port is exposed to vacuum as the throttle plate moves past the port—usually slightly off-idle.

Spark-advance vacuum versus airflow calibration is closely established in manufacturing because this dramatically affects HC and NO_x emissions. Distributor advance, once considered so important for economy, is an essential link in the emission-reduction chain.

VALVE TIMING

Valve timing has the greatest effect on an engine's idling and low-speed performance. A racing cam adds valve overlap and lift. While this allows the engine to breathe better at high rpm, it lowers manifold vacuum at idle and low speeds. Distribution and vaporization problems become obvious as the engine becomes hard to start, idles roughly (or not at all), and has a bad flat spot coming off idle. Very poor pulling power (torque) at low rpm is another characteristic. This is especially true when a racing cam is teamed with a lean, emission-type carburetor.

Because manifold vacuum is reduced, the signal available to pull mixture through the idle system is also reduced, and the mixture is leaned. Also, the throttle has to be opened farther than usual to get enough mixture into the engine for idling. This can place the off-idle slot/port in the wrong relationship to the throttle, so there is insufficient off-idle fuel to carry the engine until main-system flow begins.

And, the idle mixture has to be made richer to offset poor vaporization and distribution problems. Part of the uneven distribution and poor driveability problem stems from the overlap period.

When exhaust and intake valves open simultaneously, some exhaust gas is still in the cylinder at higher-than-atmospheric pressure. These gases rush into the intake manifold to dilute the incoming charge. Charge dilution effectively *lowers* the combustion pressure. This is especially true up to the rpm where the overlap time interval becomes short enough so the *reverse pulsing* becomes insignificant.

Reverse pulsing through the venturis at WOT adds to the fuel flow, making a richer mixture. Once the main system starts, the discharge nozzle delivers fuel in response to airflow in either direction. More information about valve timing and reverse pulsing is on page 118.

When manifold vacuum is reduced (pressure increases toward atmospheric), the power valve may start to operate. Or it may flutter open and close as manifold vacuum varies wildly. These aren't valid reasons to remove the power valve, but do select a power valve that will be closed at the lowest vacuum (highest pressure) during idling. Use a power valve with a lower vacuum rating, i.e., a smaller number stamped on the valve.

In extreme cases, a wild racing cam magnifies these problems so the car becomes undriveable for anything except competition. This is especially true if carburetion capacity is increased to match the cam's deep-breathing characteristics.

When a racing cam is installed at the same time the carburetor is changed, the carburetor gets blamed for poor idling. The real culprit is the racing camshaft. Information on engine tuning with a racing cam is on page 131.

TEMPERATURE

Temperature affects carburetion. It affects mixture ratio because air becomes less dense as temperature increases (approximately 1% for every 11F). The density change reduces VE and power, even though main-jet corrections of approximately one main-jet size smaller for every 40F ambient temperature increase will keep mixture ratio correct.

ICE CAN CAUSE DAMAGE

The accelerator pump is the lowest point in the fuel inlet system of a Holley with removable fuel bowls or a 2010, 4010 or 4011. Water can collect there, particularly in a carburetor that isn't used regularly. If this water freezes, using the carburetor could rupture the pump diaphragm.

To avoid this problem, purge water from the carburetor's fuel system. Add a can of "dry gas" compound to the gas tank to absorb water in the fuel supply system so it will be carried into the engine with the fuel. This should be done at the start of the cold-weather season and anytime humidity is high.

Maximum power production requires keeping the inlet charge as cool as possible. For this reason, racing engines are designed or assembled so the intake manifold is not heated by exhaust gas. Most stock passenger car and truck engines use an exhaust-heated intake manifold because the warmer mixture, although not ideal for maximum power, helps driveability.

Supplying a warm-air inlet to the carburetor, or an exhaust hot spot in the intake manifold aids vaporization. Good vaporization ensures more even mixture distribution to the cylinders because fuel vapors move more easily than liquid fuel.

Icing—This occurs most frequently at 40F (5C) and relative humidity of 90% or higher. It is typically a problem at idle: ice forms between the throttle plate and bore. It usually occurs when the car has been driven a short distance and stopped with the engine idling. Ice builds around the throttle plate and shuts off mixture flow so the engine stops. Once the engine stalls, vaporization stops and the ice promptly melts. The engine can be restarted. This may occur several times until engine temperature warms the carburetor body so vaporization doesn't cause icing.

Turnpike icing occurs in the venturi system when running at a relatively constant speed for a long period under ideal icing conditions. Ice buildup gradually chokes down the venturi size so the engine runs slower and slower.

Fuel vaporization removes plenty of heat from the surrounding parts of the carburetor. So, there is a greater tendency for this phenomenon to occur in small venturis where vaporization is best. Icing is no longer a major problem because cars are factory-equipped with exhaust-manifold stoves to warm the air supplied to the air cleaner inlet. In some applications thermostatic flapper valves shut off the hot air when underhood temperatures reach a certain level.

On some high-performance engines, a vacuum diaphragm opens the air cleaner to a hood scoop or other cold-air source at low vacuums/heavy loads.

Percolation—At the other end of the thermometer there's a phenomenon called *percolation*. It usually occurs when the engine is stopped during hot weather or after it has been run long enough to be fully warm. Engineers call this a *hot soak*.

In this case, no cooling air is being blown over the engine by the fan or vehicle motion. Heat stored in the engine block and exhaust manifolds is radiated and conducted directly into the carburetor, fuel lines and fuel pump.

Fuel in the main system between the fuel bowl and main discharge nozzle can boil or percolate. Vapor bubbles push or lift liquid fuel out of the main system into the venturi. The action is similar to that in a percolator coffee pot.

Fuel falls onto the throttle plate and trickles into the manifold. Excess vapors from the fuel bowl, and from the bubbles escaping from the main well, are heavier than air and drift down into the manifold. This makes the engine difficult to start and a long cranking period is required. In severe cases, enough fuel collects in the manifold so it runs into cylinders with open intake valves. Fuel washes oil off the cylinder walls and rings, causing excessive engine wear.

Percolation is aggravated by fuel boiling in the fuel pump and in the fuel line to the carburetor. Because this creates fuel pressure as high as 15–18 psi, the inlet valve needle may be forced off its seat. Fuel vapor and liquid fuel are forced into the bowl, raising the fuel level. This makes it that much easier for the vapor bubbles to lift fuel to the spillover point.

Solving percolation problems requires a systematic approach. The main system is designed so vapor bubbles lifting fuel toward the discharge nozzle tend to break before they can push fuel out of the nozzle. Fuel levels are carefully established to provide as much lift as can be tolerated. In some instances, the fuel must pass through an enlarged section at the top of the main well or standpipe to discourage vapor-induced spillover.

Gaskets and insulating spacers are used between the manifold and carburetor, and between the carburetor base (throttle body) and fuel bowl. An aluminum heat deflector or shield keeps some engine heat away from the carburetor.

Hot-starting problems are reduced by internal bleeds in the fuel pump. A vapor-return line may also be used on the carburetor ahead of the inlet valve. When the bleed and return line are used, any pressure buildup in the fuel line escapes harmlessly into the fuel tank or fuel-supply line.

Another bad high-temperature effect is boiling fuel in the fuel line between the pump and the fuel tank. Fuel can even boil in the fuel pump. When the fuel pump and line are filled with hot fuel, the pump supplies a mixture of vapor and liquid fuel to the carburetor.

Very little liquid fuel is delivered during an acceleration after a hot soak, so the fuel level drops, causing leaning. The bowl may be nearly emptied, partially exposing the jets. When the jets are partially exposed, the carburetor can't meter a correct fuel/air mixture because the jets are designed to work with liquid fuel—not a combination of liquid and vapor.

This condition is called *vapor lock.* Holley carburetors are especially resistant to the problem. Their large bowl holds enough liquid fuel so the engine can run even though some has escaped as vapor.

In difficult cases, fuel lines may have to be rerouted to keep them away from extreme heat, such as the exhaust system. If the lines cannot be relocated, it is usually possible to insulate them. This is especially important for racing vehicles. Cool cans (page 122) help, as do high-performance electric fuel pumps located at the tank to push fuel to the carburetor.

AIR DENSITY

In the air requirements section of Chapter 1, *Engine Requirements,* we related VE of the engine to the density of the fuel/air mixture received by its cylinders. We showed that the higher the density, the higher the VE.

The mixture density depends on atmospheric pressure, which varies with altitude, temperature and weather conditions. And, mixture density is also affected by intake-system layout. Density increases when the carburetor is supplied with cool air and when the intake manifold is not heated.

Density is reduced if the inlet air is heated or if the fuel/air mixture delivered by the carburetor is heated as it enters the manifold. Further density reductions occur as the mixture picks up heat from the manifold and cylinder head passages, hot valves, cylinder walls and piston heads.

As with almost all other engine variables, there are trade-offs. Warming the mixture reduces its density, but also improves distribution, especially at part-throttle.

Density has an effect on carburetor capacity and mixture. The *major* density changes that occur are due to altitude changes. Let's consider the effects of driving from sea level at Los Angeles to 5000 ft. at Denver. The flow and pressure-difference expressions look like this:

$$Q \sim A \sqrt{\frac{\Delta_p}{\gamma}} \quad \text{or} \quad \Delta_p \sim \left(\frac{Q}{A}\right)^2 \times \gamma$$

where

Q	=	volume flow
γ	=	density
Δ_p	=	pressure difference
A	=	carburetor-venturi area

Regardless of density, the volume taken in remains the same at a given rpm, but going from sea level to 5000 ft. drops density to 83% of its sea-level value. Pressure difference or Δ_p is also 83% of the original value, so the carburetor acts as if it were larger.

Because Δ_p increases inversely as the square of the area, area must be reduced only by a ratio of the square root of 0.83 or 0.91 to restore the same pressure difference with the original carburetor at sea level. In other words, the carburetor acts as if it were 9% larger. This is only a problem if carburetor size was marginally too big to begin with.

Consider a carburetor on the verge of stumbling and having flat spots in acceleration at sea level due to late turn-on of the main system. These problems will worsen at high altitudes because the carb size effectively increases due to reduced air density.

WOT fuel/air ratio is determined mainly by venturi size and main jet size, assuming fuel and air density never change.

A rule of thumb is to reduce Holley jet size by one number for each 2000-ft. altitude increase. Holley engineers usually figure approximately 4% fuel flow change between jet sizes. Thus, two jet sizes smaller will keep nearly equivalent fuel/air ratios at an altitude of 5000 feet.

COMPRESSION RATIO (CR)

High compression improves engine performance by increasing the burning rate of the fuel/air mixture. Peak pressures and peak torque can approach the maximum of which the engine is capable. High compression also increases HC and NO_x emissions.

Lowering the compression reduces HC emissions by reducing the combustion chamber's surface-to-volume ratio. The greater the surface-to-volume ratio, the more surface cooling occurs, thereby increasing HC.

Until about 1970, high-compression engines with up to 11:1 CR were available in high-performance cars. By 1971, manufacturers were reducing compression ratios and by 1972 most cars had no more than 8.0–8.5:1.

Reducing compression slows the burning rate of the fuel/air mixture so peak pressures that encourage NO_x formation aren't reached. Reduced compression also increases heat transfer into the cylinder walls. Burning continues as the piston is descending, thereby raising the exhaust temperature.

Low compression increases the fuel/air requirement at idle. More residual exhaust gas remains in the clearance volume and combustion chamber when the intake valve opens, causing excessive mixture dilution. This can cause off-idle driveability problems. In effect, low compression promotes some EGR without emissions plumbing or hardware.

With the advent of the three-way catalytic converters, microprocessor control of engines and higher octane unleaded fuels, engineers were able to tune up engines. Compression ratios were again increased so that by 1986, 9.0–10.0:1 CRs were fairly common in production cars.

Raising or lowering the compression ratio of an engine doesn't normally affect the main system fuel requirements, so jet changes aren't usually required. Raising compression may require slightly less ignition advance in some cases.

CR and Fuel Octane—Compression ratio and octane requirements are closely related. As compression is increased, octane must also be increased. A higher octane fuel is required to avoid detonation and preignition—often called *knock*.

After World War II, high-compression engines were designed and produced to obtain higher efficiencies. By 1969, some had 11:1 compression!

Then the requirement for reduced emissions began to be tackled in earnest. First, the automakers asked the fuel companies to start "getting the lead out" for emission equipment that wouldn't be able to tolerate lead in the exhaust.

Engineers created engines to operate on lead-free gasoline. Tetraethyl lead (Ethyl compound) was one of the most commonly used anti-knock additives used as an octane-increaser in gasolines. Lead compounds affect plant life, the atmosphere and humans. Consequently, there are serious health and ecological concerns about lead content in car fuels.

But, these weren't the only reasons engineers wanted to eliminate lead from fuel. The expensive catalyst used in converters (standard parts from 1975 on) is destroyed when contaminated with lead and lead byproducts.

The engineers, along with the oil companies, designed a small filler orifice for the gas tank. The unleaded gas pumps were equipped with nozzles to fit. This was an attempt to ensure that drivers couldn't use gasoline that could wreck the catalytic converters.

At the start of the measures, unleaded regular was 87 octane (pump reading). Premium unleaded was 91 octane (pump reading), but it was difficult to find. Premium unleaded of 92 octane (pump reading) is readily available in the '90s. Gasoline refineries have steadily reduced octane ratings and the trend is to still lower ones. Although 100+ octane gasolines were commonplace in the late '60s, 87 octane unleaded regular is the primary fuel as of 1994.

EXHAUST BACK PRESSURE

Exhaust back pressure has little effect on carburetion at low speeds. For high performance, low back pressure is desired to obtain the best volumetric efficiency through optimum breathing. The effects of exhaust restriction increase approximately as the square of rpm.

Using headers may change the main-jet requirement. Either richer or leaner main jets may be needed, depending on the interrelation of the engine components.

DIESELING OR "RUN ON"
This is the tendency of an engine to continue running irregularly and roughly after the ignition has been switched off. This problem is aggravated by:

- Anything that remains hot enough to ignite fuel/air mixtures—such as any sharp edges in the combustion chamber.
- High idle-speed settings used to meet emission requirements.
- Combustion chamber deposits.
- Low-octane fuel.

Performance Tuning 8

Butch Leal's Beretta lined up for a qualifying run. All Pro Stocks use two Holley 4500 carburetors. NHRA 1991 Worldfinals photo by Michael Lutfy.

The variables covered in the previous chapter also affect performance, so be sure to read both chapters.

RAM TUNING

Ram tuning can give better cylinder filling and improved VE in a narrow speed range. A combination of engine-design features are involved:

• Intake and/or exhaust-system passages or pipe lengths.
• Valve timing.
• Velocity of intake and exhaust gases.

Ram tuning improves torque at one point or narrow rpm band, but the improvement tends to be "peaky." Power falls off sharply on either side of the peak. Ram tuning is a resonance phenomenon sought at a tuned peak with the knowledge that *the power gained at that point may be offset by corresponding losses at other speeds.*

While ram tuning can add mid-range torque, these gains are obtained at the expense of top end power. Or, more usual for high-performance engines, low- and mid-range torque may be sacrificed to take advantage of top-end improvements.

Individual intake-manifold passages for each cylinder (ram or tuned length) can be measured from the intake valve to the entry of a plenum chamber fed by one or more carburetors, or to the carburetor venturi.

If venturi size is significantly smaller than the passage *between* the venturi and the valve, the venturi forms a *reflection point* and defines tuned pipe length.

If the venturi is about the same size as the passage between the venturi and the valve, then it does not form a reflection point. The tuned pipe length is determined by the next large change in section outward—probably the air horn or carburetor inlet.

Engine speed at which ram effect is most pronounced varies inversely with tuned length. Therefore, the shorter the tuned length, the higher the rpm at which peak torque will occur. Conversely, the longer the passage length, the lower the rpm at which peak torque occurs.

Equations to describe where ram-tuning effects will occur are over-simplifications because they leave out the effects of manifold-passage size and the sizes of intake ports and valves. Making any of these larger raises the rpm at which best filling occurs. This explains why best driveability and street performance is

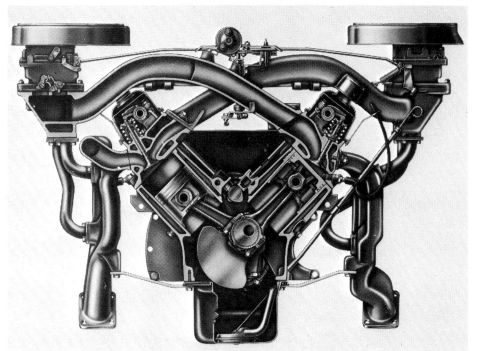

1963 Chrysler 413-cid 300J Ram Induction engine had eight equal-length intake passages. Each four-cylinder set was fed by one carburetor. An equalizing tube connected the two sets. Runner lengths were selected to give 10% torque improvement at 2800 rpm. The strong torque increase provided noticeable acceleration improvement in a 1500-rpm range from 50–80 mph. This was a classic example of ram tuning in a production car.

obtained with small-port manifolds and heads.

Torque peak can be modified by changing valve lift and overlap and by adding a plenum chamber between the runners and the carburetors. Using a plenum under the carburetors lowers the torque peak that can be achieved, and also broadens the torque over a wider rpm range. Thus, the engine becomes less *peaky* and easier to tune.

Installing a plenum also reduces fuel standoff and allows each cylinder to draw additional mixture from the other carburetor barrels at high rpm. This greatly reduces the carburetor airflow-capacity requirement for the engine.

Valve timing greatly affects rpm capabilities. Cylinder filling is aided at high rpm by holding the intake valve open past BC (Bottom Center). At low speeds, holding the valve open past BC allows part of the intake charge to be blown back onto the intake manifold as the piston rises on

its compression stroke. This reversed charge reduces manifold vacuum and drastically affects idle and off-idle F/A mixture requirements.

As rpm increases, faster piston movement creates a greater pressure drop across the carburetor. Air enters the carburetor with higher velocity, giving greater acceleration (and momentum) to the mixture traveling toward the valve. Thus, as the piston approaches BC on the intake stroke, cylinder pressure is rising toward that at the intake port. And, pressure at the intake port is being increased by air-column momentum in the intake-manifold passage supplying it. Thus, filling improves with rpm until friction losses in the manifold exceed the gain obtained from delayed valve closing.

The past two paragraphs are true, regardless of whether ram tuning is used or not. Now let's consider what happens in the manifold passage as the intake valve is opened and closed.

When the piston starts down on its intake stroke, a *rarefaction* or negative-pressure pulse is reflected to the carburetor inlet.

As this pulse leaves the carburetor, atmospheric pressure rushes in as a positive-pressure pulse. When the passage length is optimum for the rpm at which peak torque is sought, the positive-pressure pulse arrives near the time when the valve is closing. The pulse assists in the last part of cylinder filling. Note the interrelationships of passage length, mixture velocity, valve timing and rpm.

The mixture attains a velocity of up to 300 ft per second or more (depending on rpm) as it travels through the port during the intake stroke. Because the mixture has mass (weight), it also has momentum. This is useful for aiding cylinder filling when the intake valve is held open after BC (sometimes to 100° past BC) while the piston is rising on the compression stroke.

When the intake valve shuts, incoming mixture piles up or *stagnates* at the valve backside, reflecting a compression (positive-pressure) pulse or wave toward the plenum or the carburetor inlet. As this wave leaves the carburetor inlet, it is followed by a negative-pressure pulse back to the valve. This bouncing or reflective phenomenon repeats *several times* until the inlet valve again opens. Pressure at the carburetor inlet varies from positive to negative as the wave bounces back and forth in the inlet passage.

At certain engine speeds, the reflection or resonance phenomenon will tend to be in phase (synchronized) with intake valve opening and closing. The positive pulses will tend to "ram" the mixture into the cylinder. Improved filling is the result.

Although the pulsations are in phase only at certain speeds (yes, there are multiple peaks!), the mixture column in the intake passage provides some ram effect at all speeds because

Chrysler's production plenum-ram for Hemi engine was constructed to provide peak power around 6800 rpm. Branch length and plenum volume were varied to fit specific racing applications. Large-block wedge engines had a similar manifold. Holley offers 0-4235 and 0-4236 770 cfm Model 4160s as replacements for the original List 3116 vacuum secondary carbs with 1-11/16-in. throttle bores.

of its own inertia. Best filling is obtained with intake-system and exhaust-system resonances in phase at the same rpm.

On the exhaust side, exhaust gas enters the pipe at 80 psi or higher because the exhaust valve opens before BC (before the power stroke ends) while there is still pressure in the cylinder. Thus, the exhaust gets a "head start" so the piston does not have to work so hard pushing out the exhaust. The exhaust "pulse" starts a pressure wave traveling at the speed of sound (in hot gas) to the end of the system. From the end of the system, a rarefaction or low/negative-pressure pulse reflects back to the exhaust valve at the same speed.

Tuned systems are constructed with lengths to allow this pulse to arrive during the overlap period. The idea is to reduce exhaust residuals in the clearance volume and ensure complete emptying of the cylinder. This reduces charge dilution and provides

more volume for F/A mixture—hence greater volumetric efficiency.

Main Jet Requirements–Ram tuning has a dramatic effect on main-jet requirements because strong pulsing at WOT pulls fuel out of the discharge nozzle in *both* directions. If you're concerned about part-throttle performance, the F/A mixture must be richened. Adding a plenum chamber to the manifold or carburetor base softens high-rpm WOT pulsing and allows you to use a compromise jetting. This usually handles both part-throttle and WOT operation.

Tuned-exhaust effects on the main-jet requirement vary. If the exhaust causes stronger intake system pulsing, a smaller main jet could be used; if the effect lessens pulsing, a larger jet may be required. Predicting these effects is impossible, so tuners tackle each situation by trial and error.

Summary of Ram Tuning–Peaky results obtained with ram tuning seriously reduce engine flexibility.

RAM AIR

The forward motion of your car can induce some air pressure into the intake system, provided a forward-facing inlet is connected to the carburetor.

While very minor pressure increases are obtained, even a minor pressure aids induction at high rpm and gives more HP where the engine is starting to "run out of breath." According to Gary Knutsen of McLaren Engines, pressures of 6 in. of water (about 0.22 psi) were obtained at racing speeds on the Chaparral race cars. On very long straights this pressure provided measurable performance benefits.

Other writers have claimed improvements of as much as +1.2% at 100 mph, +2.7% at 150 mph and +4.8% at 300 mph. Such increases can make the difference between winning and losing.

Air Scoops–You may have noted some race cars equipped with a forward-facing air scoop over the induction system. These typically have the scoop opening *ahead* of the hood. It should be far enough *above* the hood or roll bar so the scoop picks up relatively unturbulent air.

If air scoops are used to duct air to the carburetor, do not connect the scoop directly to the carburetor unless it includes air-straightening devices. Instead, connect the scoop or duct to a cold-air box or to the air-cleaner housing to avoid creating turbulence as the incoming air enters the carburetor air horn.

The scoop should mate with a tray under the carburetor/s so any ram-air pressure is not lost into the engine compartment and to keep warm underhood air out of the carburetors.

If a tall manifold such as a plenum-ram is used, the hood must be cut for clearance and a scoop added to cover the carburetor/s. Whether the scoop opens at the front or back depends on airflow over the car. The optimum air entry into the scoop may have to be determined by testing. An optimum entry provides an above-atmospheric-pressure air supply that is non-turbulent.

In general, scoop-opening area should be approximately 12% larger than carburetor-venturi area. The scoop roof should be positioned 1-1/2-in. above the carburetor inlet. Any more clearance may create detrimen-

tal turbulence and any less restricts airflow into the carburetor.

COLD AIR & DENSITY

Mixture density has been thoroughly discussed in Chapter 1, *Engine Requirements*. There we show that higher density inlet air improves the engine's VE proportionate to the density increase. So, let's examine the practical aspects. What can you do to keep density "up" to get the best horsepower from your engine?

First, underhood temperature is not ideal for hp production. Even on a reasonably cool day, air reaching the carburetor inlet has been warmed by passing through the radiator and over the hot engine components. Underhood temperatures soar to 175F (80C) and higher when the engine is turned off and the car stands in the sun. An engine ingesting warm air loses more power than you might imagine. Assume that the outside (ambient) air temperature is 70F (21C) and the underhood temperature is 150F (66C). Use the following equation:

$$\gamma oa = \sqrt{\frac{460 + t_{uh}}{460 + t_{oa}}} \times \gamma uha$$

$$\gamma oa = \sqrt{\frac{460 + 150}{460 + 70}} \times \gamma uha$$

$$\gamma oa = \sqrt{1.15} \; (\gamma uha) \text{ or } 1.072 \; \gamma uha$$

where

γoa = outside air density
γuha = underhood air density
t_{oa} = outside air temperature
t_{uh} = underhood air temperature

In this example, outside-air density is 107.2% of underhood-air density, or 7.2% greater. Because mass airflow increases in direct proportion with density, hp with outside air will increase at the square root of 1.15, which is 1.072 or 7.2%. If the engine produces 300 hp with 150F (66C) air-inlet temperature, it can be expected to produce 322 hp with 70F (21C) air-inlet temperature. The density

increase caused by using outside air is considerable.

Cold air gives more improvement than ram air because approximately 1% hp increase is gained for every 11F drop in temperature. This assumes the mixture is adjusted to compensate for the density change and there is no detonation or other problems. Using outside air instead of underhood air is climate-limited because too-cold temperatures may cause carburetor icing.

Drag racers always keep the hood open and avoid running the engine between events so the compartment stays as cool as possible. It is helpful to spray water onto the radiator to help cool it. This ensures the engine water temperature will be lower and the radiator will not heat incoming air any more than is necessary.

Although "seat-of-the-pants" feel may indicate stronger performance from a cold engine, the fact of the matter is that the engine coolant temperature should be around 180F (82C) or so to allow minimum friction inside the engine. Keep the engine oil and water temperature at operating levels while taking care to keep down the inlet-air temperature.

If you use a cold-air kit that picks up cold air at the front bumper, change the air-cleaner filter element at frequent intervals—perhaps as often as once a week in dustier areas. If you leave the filter out of the system, your engine will quickly wear out.

You are better off ducting cold air from the cowl just ahead of the windshield. This high-pressure area ensures a supply of cool outside air to the carburetor. That area still gets airborne dust, but it is several feet off of the ground and away from some of the heavier grimy grit encountered at road level. Using fresh air from the cowl is another way to get performance while keeping your car looking stock.

Cars with stock hoods and stock or near-stock-height manifolds may be

You can calculate relative air density from barometer and temperature readings, but most racers prefer an air density meter. It can be used to help select correct main-jet area according to square root of air-density changes in percent. Air density changes from hour-to-hour and day-to-day, and most certainly from one week to the next and one altitude to another.

Pro Stock car with scoop. Base of scoop seals to carburetor inlets so ram air can be used. 1991 photo by Michael Lutfy.

equipped with fresh-air ducting to the air cleaner by using parts from some high-performance cars.

Or, you may prefer to use one of the fresh-air hoods offered on some models. These typically mate a scoop structure on the hood with the air-cleaner tray on the carburetor/s. Thus, underhood air is kept out of the carburetor.

While we are talking about improving density by using cold inlet air, let's remember that a heated manifold reduces density. Exhaust heat to the manifold should be blocked off to create a "cold" manifold for performance.

There are various ways to do this. In some instances intake-manifold gaskets are available to close off the heat openings. Or, insert a piece of stainless steel or tin-can between the gasket and manifold to block off the opening. Many competition manifolds have no heat riser and therefore the manifolds are "cold" to start with. Some car makers offer shields that fit under a V8 manifold to reduce hot oil heating of the manifold.

AIR CLEANERS

Because every engine needs an air cleaner to reduce expensive cylinder wear caused by dust, use one that will not restrict the carburetor's airflow capabilities. Avoiding restrictions allows your engine to develop full power.

The only time an engine might possibly be run without an air cleaner is on engines being operated where there is no dust in the air or pits, such as on ocean racing boats. Even then, when an air cleaner is removed from the carburetor, the air cleaner base or something similar should be retained because its shape may provide a efficient entry path for the incoming air. It also will help to keep incoming air from being heated by the engine.

The air cleaner directs air into the carburetor so the vents work correctly and air gets into the air bleeds correctly. One dyno test series showed a 3 hp loss by removing the air-cleaner base. It causes air to flow into the carburetor with less turbulence. It is essential!

In general, a tall, open-element air cleaner provides the least restriction. It also increases air-inlet noise.

Some air cleaners allow full airflow capability. These air cleaners should be used by racers, even if their use requires adding a hood "bump." An air cleaner that gives full-flow capability to the carburetor provides

Econo dragster has velocity stack sealed to base of scoop. Carburetor is on a 2-in. spacer to straighten mixture flow and improve distribution.

impressive top-end power improvements, as compared with one that restricts flow. For instance, the use of two high-performance Chevrolet air cleaners stacked together (instead of one open-element cleaner) improved a 1969 Trans Am Camaro's lap times at Donnybrook, Minnesota, by one full second.

It is very important to check clearance between the upper lid of the air cleaner and the top of the carburetor's vent tubes. Air-cleaner elements vary

Fresh-air hood components on Chevrolet Camaro. Chevrolet engineer Gerry Thompson holds air cleaner base that fits onto two four-barrel Holleys. Hood is open at back by cowl. Hood duct mates with foam-rubber gasket.

Chevy high-rise manifold for Z-28 (original 1969 LT-1 engine) had oil shield under heat riser area. Shield and blocked heat riser help keep intake charge cool.

Fuel-cooling can (cool can) is often used by drag racers. Fuel passing through coiled line is cooled by ice or dry ice and alcohol in can. Cooler fuel temperature ensures that carburetors receive liquid fuel. Cold fuel under pressure is not likely to flash into vapor when it drops to atmospheric pressure as it enters the carburetor bowl.

INCREASED DENSITY EQUALS LARGER JETS
No matter how a density increase is obtained—by increased atmospheric pressure, a cold manifold or a cooler inlet-air temperature, it most be accompanied with larger main jets. Area size increase is directly proportional to the square root of density increase in percent.

as much as 1/8 in. in height due to production tolerances. Shorter elements can place the lid too close to the vent tubes so correct bowl reference pressures are not developed. Whether the pitot tubes are angled or flat on top, there should always be at least 3/8-in. clearance between the tube tip and the underside of the air-cleaner lid.

The long snorkle intakes on modern air cleaners are intake-noise reducers, *not* performance improvers. High-hp engines nearly always have two snorkles for more air and perhaps for image, too. For competition, remove the snorkles at the cleaner housing and cut additional holes into the cleaner housing to approximate an open-element configuration. Or, expose more of the element surface by inverting the cleaner top. An open-element design is least restrictive.

When it's time to race, use a clean filter element. Keep the air cleaner base on the carburetor if you possibly can, even if you have removed the air cleaner cover and element. Secure the cleaner and/or base so it cannot vibrate off to strike the fan, radiator or distributor.

Holley found a 2.4% flow improvement when 3-1/2-in. velocity stacks were added to a 1050 cfm Model 4500. Velocity stacks clean up air entry so there is less flow-robbing turbulence.

K & N 14-in.-dia. air cleaner uses 5-in.-tall reusable oiled element as shown here on Model 4150 alcohol carburetor. It can be cleaned and reused. This type of filter is popular for off-road and dirt-track competition.

K & N Stub Stack is another air-entry device for Holley two- or four-barrel carburetors. Flow improvements of several percent are claimed. Device can be used with or without an air cleaner.

Because the carburetor is internally balanced, that is, the vents are located in the air-horn area, no jet change is required when the air cleaner is removed.

For drag racing, air cleaners are generally not used. For any other competition where dust is involved, use a low-restriction paper-element cleaner, such as a tall open-element type and replace it often.

If the cleaner is removed, the engine can draw in a lot of abrasive dirt by merely running back down the return road. If the engine has to last, then stop at the turn off and put the air cleaner on, or push the car back to the pits.

Combination paper-element and oiled-foam cleaners should be used for very dusty conditions.

Air cleaners also protect against fires caused by starting "belch-backs," reduce intake noise, and reduce engine wear. Intake noise can be horrendous, even worse than exhaust noise, on an engine turning a lot of rpm. Air cleaners greatly reduce that noise.

Before leaving the subject of air cleaners, look at how the stock cleaner is designed before you invest in some flat-top, flat-bottom, short cleaner because it looks good. Note that the high-performance cleaner

stands high above the carburetor air inlet. Adequate space allows the incoming air to enter correctly and with minimum turbulence. With a flat filter sitting right on top of the air inlet you may lose horsepower.

VELOCITY STACKS

Velocity stacks are often seen on racing engines. These can improve cylinder filling (charging) to a certain extent, depending on many other factors.

Velocity stacks provide a straightening effect to the entering air. And, they can contain fuel standoff, which is typical with small-plenum manifolds. Stacks need space above them to allow air to enter smoothly. Mounting a hood or air-box structure too close to the top of the stacks reduces airflow into the carburetor. 1-1/2 inches is a minimum clearance between the top of a velocity stack and any structure over it.

When an air cleaner is used on a carburetor equipped with stacks, keep the 1-1/2 in. recommended clearance between the top of the stacks and the underside of the air-cleaner lid. This may require using two open-element air cleaners fastened together with RTV or another sealant—and a longer

stud between the carburetor top and the cleaner lid.

Using velocity stacks on the Model 4500 may gain some minor airflow improvement. On a four-barrel with the 5-in. air cleaner base, wide-mouthed entry devices may provide improvements ranging from less than 1% to as much as 4%.

DISTRIBUTION CHECKING

Distribution checking determines whether all cylinders are receiving an equal mixture. This is extremely important when a manifold or carburetor change is made. A previous carburetor/manifold combination may have provided nearly perfect distribution, but you cannot take the chance that one or more cylinders will be running lean. Several ways of checking distribution are detailed in the distribution section of Chapter 1, *Engine Requirements.*

Exhaust Temperature—If you have a Superflow or other computer-controlled dynamometer, an exhaust temperature printout for all cylinders will help you check distribution. Strive for differences of not more than 100F (38C), and preferably less. Fuel-injection expert Bill Howell suggests a

Dual-snorkles or inlets on air cleaner are often trademark of factory high-performance vehicles. Here is 1972 Pontiac GTO.

Chevrolet Z-28 air cleaner is one of the least-restrictive single-element air cleaners ever offered.

temperature range of 1450–1550F (788–844C) for the Chevrolet V6 or V8 small block. Other engines may require slightly different exhaust temperatures for peak power output.

Reading Spark Plugs—At the race track there's only one way to do it. You have to rely on the appearance and color of the plug electrodes and porcelains. These can provide a lot of valuable information about what is happening in the engine.

The part of most interest is the base of the porcelain. Because it is "buried" in the plug shell, an illuminated magnifier should be in your tool box.

Checking plug color gives only a rough idea of what is occurring in the way of mixture ratio and distribution. It is difficult to see a change in plug color without changing the main jet at least four sizes.

Many items cause plug-appearance variations. Using plug color to check distribution is only helpful when the engine is in good condition. Engine condition can be checked quickly with a compression gauge to make sure all cylinders provide equal compression at cranking speed. For a more accurate check, a leakdown test can compare cylinder condition.

New plugs take time to "color"— even three or four drag-strip runs may not "color" new plugs. Plug color is only meaningful when the engine is declutched and shut off at the end of a high-speed full-throttle, high-gear run. If you allow the engine to slow with the engine still running, plug appearance will be meaningless. Plug readings can be made after full-throttle runs on a chassis dyno with the transmission in an intermediate gear so the dyno is not overspeeded. But road tests require high gear to load the engine correctly.

Plug checks can be useful where the engine has been running at full throttle against full load applied by an engine dyno. It is easier to get good plug readings on the dyno because full power can be applied and the engine shut off without difficulty. However, plug heat range and carburetor jetting established on an engine dyno may not be absolutely right for the same engine installed in your race car chassis. Airflow conditions past the carburetors can easily change the requirements—perhaps so unevenly that different cylinders will need different changes.

It would be nice if every plug removed from an engine looked like the others from the same engine—in color and condition. But this is seldom ever achieved! Color and other differences indicate combustion-chamber temperatures and/or fuel/air ratios are not the same in every cylinder—or that related engine components need attention. The problem is greatly complicated in engines where there is a great difference between the cylinders in turbulence and efficiency. The big-block Chevrolet is a notable example.

If differences exist in the firing end of the plugs when you examine them, the cause may be due to:

• Unequal distribution of the mixture.
• Unequal valve timing (due to incorrect lash or a worn cam).
• Poor oil control (rings, excess clearance or valve-stem seals).

Ignition Checking—Problems within the ignition system can also lead to plugs not reading the same or misfiring:

• Loose point plate.
• Arcing in the distributor cap.
• Defective rotor, cap or plug wires/connectors.
• Cross-fire between plug wires.
• Defective primary wire or even a resistor that opens intermittently.

Pay special attention to the cleanliness of the entire ignition system including the inside and outside of the distributor cap and the outside of the coil tower. Also, clean the inside of the coil and cap cable receptacles.

If the cylinders have equal compression and the valves are lashed correctly, a difference in plug appearance may indicate a mixture-distribution problem. It may be possible to remedy this with main-jet changes.

For instance, if one or more plugs show a rich condition, install smaller main jets in the throttle bore/s feeding those cylinders.

The real problem occurs when several cylinders fed from the *same* throttle bore show different mixture conditions: some lean with some correct, or some rich with some correct, or perhaps a combination of all three conditions! This reveals a manifold fault that can't be corrected with jet changes. Correcting such conditions requires manifold rework beyond the scope of this book.

HEADER EFFECTS vs. JETTING

Headers usually reduce exhaust back pressure so the engine's VE is increased–it breathes easier! The main effects of headers are seen at WOT and high rpm.

Using headers may change the main-jet requirement. Either richer or leaner jets may be needed, depending on the interrelation of the engine components.

CARBURETOR RESTRICTION

Carburetors are tested at a given pressure drop at WOT to obtain a cfm rating indicative of flow capacity. One- and two-barrel carburetors are tested at 3.0-in.Hg pressure drop. Three- and four-barrel carburetors and fuel-injection throttle bodies are tested at 1.5-in.Hg pressure drop.

When you want to make comparisons, use these formulas:

$$\frac{\text{Equivalent airflow}}{\text{at 1.5 in.Hg}} = \frac{\text{cfm at 3.0 in.Hg}}{1.414}$$

Equivalent airflow = cfm at 1.5 in. Hg. X 1.414 at 3.0 in.Hg

The one- and two-barrel rating was adopted because low-performance engines typically showed WOT manifold-vacuum readings of 3.0 in.Hg. When four-barrel carburetors and high-performance engines became commonplace they were rated at 1.5-in.Hg pressure drop because of two reasons.

First, this rating was close to the WOT manifold vacuum being seen in these engines. Second, existing test equipment had been designed for smaller carburetors. Pump capacity on this expensive test equipment would only provide 1.5-in.Hg pressure drop through larger carburetors. Hence the 1.5-in. rating came about as a "happy accident."

For maximum output, it is essential to have the carburetor as large as possible–*consistent with the required operating (driving) range.*

Using more flow capacity than calculations indicate necessary for the engine can reduce inlet-system restriction. It increases VE at WOT and very high rpm. Such carburetion arrangements compete with fuel injection in terms of performance because the restrictions are minor. But note that the dual-quad installations typically used by professional drag racers are *not* capable of providing usable low- or mid-range performance. These engines typically operate in a very narrow range of 6000–8500 rpm or so.

TOOLS REQUIRED

Start with patience! You need more than the feel in the seat of your pants and the speedometer for serious tuning. Specific tools are required, but you'll especially need a patient and methodical approach to the project. You cannot be in a hurry!

A vacuum gauge, fuel-pressure gage and a stopwatch are essential. So is a tachometer. Vacuum and fuel-pressure gauges are an extra set of "eyes" to let you see what's happening inside of the engine. For serious competition on a regular basis, an air-density guage is helpful. You'll also want a 1-in. open-end wrench, preferably the MAC-141 that's specifically designed for fuel-inlet nuts on two- and four-barrel carburetors with center-hung floats. A broad-blade screwdriver can be used for jet changes.

TIMING DEVICES

Although a stopwatch can be used for some very fine tuning, you may want to consider getting your own timing device with associated photocells, or rent a portable unit so you can set up the lights at varying distances. Pro Stock racers often test acceleration over an initial 60 ft. to work out starting techniques, tire

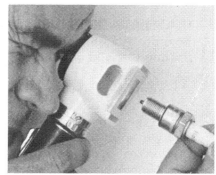

Plug porcelain color and appearance being checked with spark plug illuminator. A built-in magnifier aids plug reading. One should be in every tuner's tool kit. Engine must be turned off, declutched or put into neutral at the conclusion of a WOT full-power run in top gear. If this method is not used, plug readings are meaningless.

combinations and carburetion. Most races are won in the critical starting period and by initial acceleration over the first few feet.

A FEW PARTS WILL BE HELPFUL

When you are working on a two- or four-barrel with detachable fuel bowls, have a few parts available. Foremost among these are extra bowl and metering-block, bowl and bowl-screw gaskets. O-rings for the transfer tube, if used, should also be on your shopping list.

Before you buy main jets, find out what size is already in the carburetor. Look for yourself to be absolutely sure. Mark the jetting on the bowls. Once you know what jets you have, check Holley's *Illustrated Parts & Specs Manual* or the *Holley High-Performance Parts Catalog* to see which jets are *supposed* to be there.

Buy four sizes lean and four sizes rich for each main jet in your carburetor. This will handle most engine variations and most atmospheric changes (density).

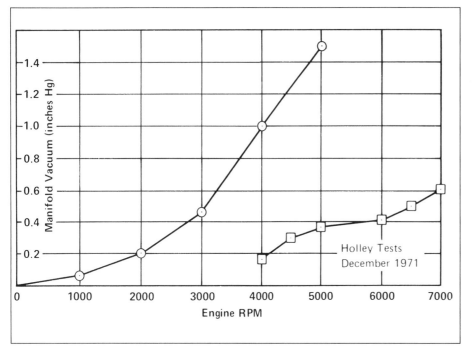

Chart shows minimum restriction by using two four-barrel carburetors for drag-race or competition engine. Such installations provide horsepower equivalent to racing fuel-injection systems. Linked circles represent Model 4165 650 cfm single carb on Chevy L-46 350-cid engine. Linked boxes are two Holley 4160 carburetors on 350-cid Chevy racing engine with plenum manifold.

Main jets are not necessarily the same size in a four-barrel. Secondaries may be different from the primary side. If you are running a big-block Chevrolet with an open-plenum manifold, the carburetor should use three different jet sizes because it is "stagger-jetted."

For plenum-type all-out drag-race manifolds, install main jets at least two sizes larger. *These are the only two cases where jets should be changed before running the engine.* Otherwise, make your first tests with the carburetor jetted *exactly as supplied by Holley.* There are no other exceptions.

For any four barrel with a 650–850 cfm main body, main jets are between 76–81, assuming power valves were originally installed in the carburetor. If the jets are more than four sizes away from the 76–81 spread, put back the original jetting before you start tuning. Again, mark the jetting on the bowls.

Drag racing often requires richer jets. For instance, if 80s worked fine on the dynamometer, you may need 82s or 84s at the strip. Jets have to be rich to get a mixture equivalent to that dialed in on the dynamometer with rock-steady conditions.

Double-pumpers or carburetors with a high-capacity pump may not need extra-rich jetting for the strip because the two accelerator pumps will usually inject enough fuel to cover up for the airflow lag.

ONE CHANGE AT A TIME

It never occurs to some would-be tuners, and even some old-timers, to make ONE CHANGE AT A TIME. It's easy to be tempted, especially when you are sure you need a heavier flywheel, different gear ratio, other tires, different main jets, two degrees more spark advance and a different plug heat-range. Changing these one at a time would be just TOO SLOW.

You convince yourself of that. And, first thing you know, you've lost the baseline tune. You won't know where you're at or how you got there.

Anytime you change more than one thing, you no longer have any idea of which change helped (or hindered) performance. It's even possible that one change provided a positive improvement that was cancelled by the negative effects of the other change. The net result *seemed to be* no change.

Check each change against known performance by the clocks, or by your stopwatch and tachometer.

When tuning, follow a procedure and stick to it. One person has to be in charge and notes must be kept for each change and its result. The less help (and therefore, advice) you have, the better. Concentrate. Be deliberate.

Don't be surprised when the process eats up more time than you ever thought it could. If you're planning other changes soon, such as a different manifold, camshaft, air scoop, distributor-advance curve, cylinder heads or exhaust system, then put off tuning until the car and engine are set up as you expect to run it.

Whatever you do, don't fall into the common trap of rejetting the carburetor to some specialized calibration that you read about in a magazine article or heard about at the drag strip last week.

When you make jet changes, use a grease pencil or a marking pen to mark the bowls with the main-jet sizes you've installed.

Quick-change jet kits include primary and secondary bowls with removable plugs for access to the main jets without removing the float bowls. A jet-removal tool holds the jet so it can be remved from or screwed into the metering block.

TAKING CARE OF JETS

Don't Drill Jets—Jets are carefully machined with a squared-off entry and a precisely maintained exit angle. Drilling a jet won't supply the same flow characteristics as a genuine Holley jet, even though drilling apparently produces the same size opening.

Carrying Jets—Store and carry jets in a jet holder. Drill and tap a piece of wood, plastic or aluminum with 1/4-32 NF threads and use that. Some tuners use plastic boxes or even small envelopes. Don't ever carry jets strung onto a wire because this damages the jets.

Correct Tool—use a screwdriver with a wide enough blade, or use a jet tool when installing jets so there is no danger of burring the slot. Burrs on the slot edges can affect main jet flow.

BEFORE YOU START TUNING

Now that you've gathered the tools, spare jets and other paraphernalia, here are a few more details. First and foremost, check the accelerator pumps to make sure they operate with the slightest movement. And, check that there is at least 0.015–0.020-in. added travel in the pump-operating lever at WOT. See photo, page 78.

Check and set fuel levels.

Remove the air cleaner (temporarily!) so you can look into the air horn. Have someone else mash the pedal to the floor as you check with a flashlight to make sure the throttles fully open (not slightly angled). If they're not opening fully, figure out why and fix the problem.

Any time you remove and replace the carburetor, check again to ensure that you have a fully opening throttle. It is the easiest item to overlook and the cause of a lot of lost races or poor times.

For racing, take the sintered-bronze filters out of the inlets to the fuel bowls. Make sure a filter is in the line between the pump and the carburetor/s. Also make sure the choke is locked open (for racing only!).

ACCELERATOR PUMP TUNING

Shooter Size—The engine's ability to come off the drag-race starting line "clean" indicates a pump "shot" adequate for the application. A common complaint is, "It won't take the gas." Actually the engine is *not getting enough gas* so a "bog" occurs.

Two symptoms may appear. First, the car bogs–then goes. The pump-discharge nozzles may be too small so not enough fuel is supplied fast enough.

Solving this problem may mean using larger pump-discharge nozzles, which may require a larger pump, too.

Tune shooter-size by increasing or decreasing the nozzle diameter until crisp response is obtained when the throttle is "winged" (snapped open) on a free engine (no load). When crisp response is obtained, increase the nozzle size another 0.002 in. The combination will probably be driveable for a drag application.

Pump Shot Duration–Pump-shot duration can be timed by shooter size, pump-cam lift and pump capacity. Shooter size also determines the rate fuel is fed from the accelerator-pump system during WOT "slams." The override spring "gives" (compresses) when the throttle is slammed open. Compressed spring force against the lever operating the diaphragm establishes delivery pressure in the pump system. Delivery rate then depends on system pressure and shooter size.

The override spring must never be adjusted so it is *coil-bound* or has no capability of being compressed. Never replace the spring with a solid bushing to "improve" pump action. Either will cause a ruptured pump diaphragm and/or a badly bent pump linkage because gasoline won't compress. Something has to give or break if the throttle is slammed open and there's no shock absorber.

The second drag-start symptom is one of the car starting off in good

127

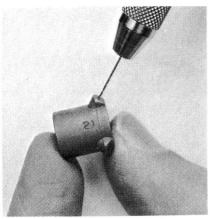

Use a pin vise to hold drill for accelerator-pump shooter. "21" stamped on shooter is original hole size in thousandths of an inch.

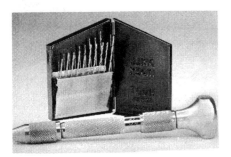

Wire drill index with drills from 00135—0.039 in. (Numbers 80—61) is useful for serious tuners. Pin vise holds drill bit that must be twisted carefully only with your fingers. These drills snap off very easily, so care is absolutely essential!

fashion—then bogging—then going once more. Here, the pump-discharge nozzles may be correctly sized, but the pump volume is not sufficient to carry the engine through until airflow is established to start the main system.

As a general rule, the more load the engine sees, the more pump shot that's needed in rate and volume.

If the engine sees less load as the vehicle leaves the line, shooter (discharge-nozzle) size can be reduced. Less load occurs when an 1800-rpm stall speed converter is replaced with a 3000-rpm converter in an automatic transmission. The same is true when replacing a light flywheel with a heavier one.

Valve timing affects pump-shot requirement. Long valve timing

(duration) and wide overlap create a need for more pump shot than a stock camshaft requires.

Carburetor size and position also affect pump-shot requirements. More pump shot is needed when the carburetor is mounted a long way from the intake ports, as on a plenum manifold.

The larger the carburetor in relation to engine displacement and rpm–the more need for a sizeable pump shot. This covers up the "hole" caused by slamming the throttles wide open. This is especially true with mechanically operated secondaries.

Holley warns against converting vacuum-operated secondary throttles to mechanical operation because there's no second pump to cover up the "hole" caused by secondary opening. Double-pumpers give an adequate pump shot to cover up mechanically opening the secondaries.

Sometimes a carburetor is changed from one engine to another or the engine is changed into a different vehicle. Or, drastic changes may be made in the vehicle itself. The main problem will always be tip-in performance. Work will be needed to get the accelerator-pump system to perform as you'd like.

If the pump cam provides full lift and therefore full travel for the pump diaphragm, cam *shape* is not important for drag-race applications. *Position* of the cam on the throttle shaft is *very important* for the drag racer. Pump cam lift is all in by about 20° throttle opening on most of the cams. Thus, a wide throttle opening can use up 40–50% of the available pump shot. If the throttle is opened

very far to provide staging rpm, so much of the cam lift is used up that the pump cannot deliver a full shot.

Determine what throttle opening is required to obtain the desired staging rpm. Then rotate the primary pump cam backward on the shaft until the pump lever again rests on the heel of the cam (no-lift position). One of the existing holes may line up with a hole in the throttle lever, or another hole may be needed in the plastic cam.

Check that there is NO clearance between the pump actuating lever and the cam. Resetting the throttle to a lower idle speed can move the cam away from the lever and delay the pump shot. A mere 2° throttle movement should move the pump lever.

When readjusting the pump-operating-lever adjusting screw to reestablish contact with the cam, check that there is 0.015–0.020 in. additional travel for the diaphragm lever at WOT between the lever and the adjusting screw.

Various cams are offered for the 2300, 4150/60/80, 4165 and 4500 carburetors. These are primarily for use in tuning the actuating of the accelerator pump/s on carburetors used for engines being driven with varying throttle openings. This includes street and highway use, road courses and circle-track racing.

Accelerator Pump Tuning Kit 36-184–includes two each of 12 pump shooters ranging from 0.025 to 0.052 in., plus two each of 10 accelerator pump cams with different profiles. The kit also includes five each of the GFLT diaphragms for 30cc and 50cc accelerator pumps.

Tuning/Calibration Kit 36-182 for Double-Pumpers–includes four each of eight main jet sizes from 70 through 84, and five pump shooter sizes from 35 through 50. Kit also includes 12 each metering block and float bowl gaskets, plus miscellaneous other gaskets.

Pump Cam Assortment Kit 20-12–includes five accelerator pump cams.

POWER-VALVE TUNING

You may immediately think, "I know just what to do–take it out!" No, regardless of all the material written to the contrary, *there is seldom any real reason to take out the power valve* and replace it with a plug.

The power valve in the secondary allows using smaller main jets. Braking forces don't cause the engine to run excessively rich and there's no tendency for the engine to load up or run rich at part-throttle.

Power valves have an important purpose—they are a "switch" between the cruising mixture ratio and full-power mixture.

Let's use as an example a car equipped with a camshaft that provides such low manifold vacuum at idle or part-throttle that the power valve opens, giving richer mixtures or perhaps flutters open and close due to vacuum fluctuations. In this instance, a power valve that opens at a still lower vacuum should be installed. If a 65 power valve is in the carburetor and the vacuum occasionally drops to 5 in.Hg at idle or part-throttle, install a 40 power valve (opens at 4 in.Hg). This ensures that the power valve will not open until its added fuel is needed.

Power-valve tuning requires using a vacuum gauge that is not highly damped, that is, the needle has to "jump" to follow vacuum fluctuations or you won't know how low the vacuum is getting. It is essential to know the manifold vacuum at idle.

You may be racing in a class

where carburetion is limited to a certain carburetor type or size that is too small for the engine. A class demanding the use of a single two-barrel is typical. Here, manifold vacuum may remain fairly high as the car is driven through the traps at the end of the quarter.

The power valve must have a higher opening point than the highest manifold vacuum attained during the run, especially at the end. If this is not the case, the power valve will close and the engine will run lean. Disaster will result–usually in the form of a holed piston. Sometimes it will get lean enough to cause "popping" sounds from the exhaust.

For example, a carburetor is equipped with a 30 (3.0 in.Hg) power valve. Testing shows that manifold vacuum is 4.0 in.Hg through the high-speed portion of the course. Change the power valve to a 65 or 85. Keep in mind the previous example so the power valve does not open at idle because of manifold-vacuum fluctuations created by camshaft characteristics.

Flat-track racing (especially with super-modified cars) and slaloms (autocross), are the only applications that generate G loads high enough to move fuel away from the power-valve inlet so air can enter to lean the mixture. If the secondary power valve is removed, main-jet size must be increased to compensate for the lost area of the power-valve channel restriction/s. Expect flooding through the main jets and out the discharge nozzles during braking or hard stops. The extra-rich mixture during part-throttle operation can cause plug fouling or loading up during "light-throttle" use such as warming up or running slowly during caution laps.

TUNING IN THE VEHICLE

A lot of engine-development work related to getting carburetion correct

> **POWER VALVE CHANNEL RESTRICTION (PCVR)**
> A large-area PVCR is sometimes used on secondaries to allow using smaller main jets for best fuel control under severe braking. If reducing PVCR diameter, close original hole with lead shot or Devcon F aluminum-base epoxy compound. Then redrill to the desired size. Or, drill the PVCR in a metering block to allow pressing in idle-feed restrictions obtained from an old metering block. Press in the brass restriction and redrill to needed size.

can be done on an engine dynamometer. However, some tuning has to be done with the engine installed. In general, it is safe to jet up (richer) one or two jet sizes when moving the engine from the dyno to the chassis.

For tuning you need a place that is always available whenever you want to use it for tuning purposes. The reason for using the same place is that subtle changes in roads can really throw off your best tuning efforts. Always use the same strip. A road can look perfectly level and yet include a substantial grade of several percent. You won't be able to tell this with your naked eye. Surveyor's apparatus is needed to make this kind of judgment.

You also need a vacuum gauge and a stopwatch.

If you are tuning for top-end performance, and you know the engine rpm at the end of the quarter mile, a lot of tuning can be accomplished with a stopwatch. Start the watch as you accelerate past an rpm point which is 2000 or 3000 below where you want to be at the top end. Stop the watch when you reach the rpm marking the end of the range in which you are interested. Use the highest

Whistle vent in top metering block ensures bowl venting and reduces possibility of fuel spewing out of bowl vents under cornering, acceleration and braking forces. It provides foam control under hot conditions, too. First whistle vents were supplied in carburetors used by Ford at LeMans. Vents are available as Part 26-40. Bent-brass baffle with triangular opening (middle block) is used in current production, usually only on primary bowls of four-barrels. It keeps foam out of bowl vents under hot conditions. Perforated baffle in lower metering block was first baffle type used. No longer used in production, it is available as Part 26-39.

gear to eliminate gear changes. Eliminating these gets rid of one more variable in the tuning procedure.

Timing runs from 3500 rpm to the peak rpm that will be used gives a very accurate indication: Is a change helping or hurting performance? High

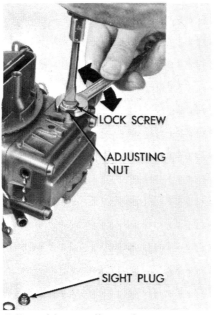

Adjustable needle and seat assemblies used on Holleys are adjusted by loosening lock screw. Turn nut to raise or lower fuel level. Locking the adjustment is the only function of the lock screw.

gear stretches the time required to pass through the rpm range of interest and eliminates wheelspin that often occurs at gear changes. You can see the effects of changes in main-jet size or pump calibration.

The race course is *not* the place to tune during actual competition. There is never enough time to get the combination running exactly right and there is always a lot of confusion in the busy, exciting and emotion-charged atmosphere of racing day. The competitor who has to do any more than fine tuning or adjustments is literally not ready to race. Further, at a drag race, conditions constantly change so times change. As more rubber is laid down at the starting line, traction improves and times get faster–*without changing the car.*

If you can use the drag strip where you ordinarily race, this is an excellent place to tune. If the timing devices are installed and operating, you'll get instant feedback on your tuning efforts. Use timing devices and the convenience of an unchanging

PULSATIONS AFFECT MAIN JET SELECTION

Because fuel flow is affected by airflow pulsations in the intake manifold, main-jet size depends somewhat on the number of cylinders fed by the venturi. Pulsing can be extreme at WOT when only one or two cylinders are fed by a single venturi. It is aggravated when no balance chamber or plenum connects the cylinder/s with the others. In a plenum manifold, reducing plenum volume *increases* pulsing. Increased pulsing *reduces* the main-jet size requirement.

stretch of road or drag strip to get things exactly right without any time pressures from competitive action.

Professional racers use on-board computers to take data during tuning and in the actual race events. A printout can be evaluated to see the effects of changes.

Standardize your starting procedure, preferably eliminating any standing starts because wheelspin at the line makes a lot of difference in your times. Wheelspin will confuse your best tuning efforts. If you want to work on your starting techniques, do that separately when you have the car tuned to your complete satisfaction.

Times or speeds that a car is turning–assuming good, consistent starting techniques–are a good indication. Is the mixture ratio being supplied by the carburetor/s correct? If a jet change makes the vehicle go faster, the change was probably made in the right direction, regardless of what spark plug color tells you. As long as a change produces improvement, keep making changes in that direction until the speed falls off. Then go back to the combination that gave the best time.

Unless plugs indicate the mixture is rich, keep richening until times start to fall off. Plug color can indicate perfect mixture, even though the engine is getting into detonation at the upper end. That's why some tuners look at the piston crowns to get an idea of what's happening in the engine.

Because the drag-race engine spends such a small portion of the run at a peak-power condition, spark plugs—especially in a quench-type (wedge) combustion chamber—may need to be bone-white for the best times. Don't be concerned about plugs not coloring in drag events so long as the times keep improving. Just keep making one change at a time–and *only one!*

When reading spark plugs, remember plugs have tolerances, too. One set of plugs may read one way. Another set with the same heat range may read differently. If you can't get plugs to read correctly, try one step colder plugs—then one range hotter. Plugs often give you a clue as to how the mixture is being distributed to the various cylinders. This is true if the engine is in good tune and the compression is the same in all cylinders—and all valves are seating.

STAGGER JETTING

If you are working with an engine that has stagger jetting, make jet changes up or down in equal increments. If one jet is normally 78 and another an 80 and you are richening by two steps, move up to an 80 in the 78 hole and an 82 in the 80 hole. Move everything equally.

If the engine has "square" jetting (same size in all four holes or same size in primary barrels with a different "same size" in the secondaries), make the changes equally for each jet position.

Underside of 0-4412 version of Model 2300 500 cfm two-barrel shows factory holes in throttle plates (arrows). These holes allow throttles to be more closed at idle so they relate correctly to idle-transfer slots. Holes are on same side of throttle as transfer slots. Light placed behind carburetor shows all around the throttles because throttles are not sealed against throttle bores. This correct clearance is established at the factory. Don't make plates fit tightly against throttle bores!

SPECIAL PROCEDURES FOR "WILD" CAMSHAFTS

The next few paragraphs are very specialized and apply only to the pro racer. Such modifications are not usually required—even with a wild camshaft. So try the carburetor in its "box-stock" condition before proceeding with such changes.

Holley performance carburetors, such as the Model 4150 double-pumpers, may be rich enough through the idle and mid-ranges to handle a wild cam without changing the idle feeds. Modifying with a hole in each primary throttle plate is common, especially on very large engines with wild cams, such as the big-block Chevrolet with a long-duration camshaft.

A wild racing camshaft with lots of valve-timing overlap can cause seemingly insurmountable

Throttle/Transfer Slot Relation

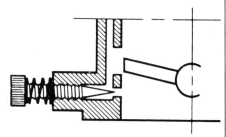

A—Transfer slot in correct relation to the throttle plate. Only a small portion of the slot opens below the plate.

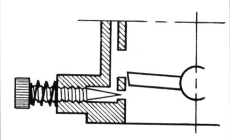

B—Throttle plate closing off slot gives smooth idle with an off-idle flat spot. Manifold vacuum starts transfer fuel flow too late. Cure requires chamfering throttle plate underside to expose slot as in **A**.

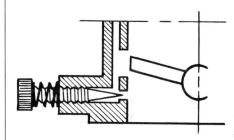

C—If slot opens 0.040 inch or more below throttle, rough idle occurs due to excessive richness. Little transfer flow occurs when throttle is opened. A long flat spot results.

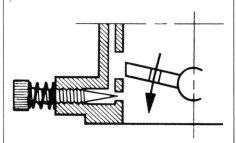

D—Correcting condition in **C** requires resetting throttle to correct position **A** and adding hole in throttle as described in text.

Two idle-feed restrictions. At left is restriction in lower end of idle fuel well. One of these pressed-in restrictions has been placed on a drill inserted in the actual restriction. "Chrysler" type metering block at right is also used on Model 4500 0-6214 and 0-6464. It has idle-fuel restriction in bottom of tube inserted into main well. One of these idle-feed restriction tubes has been laid atop metering block to show relative location. This tube is the idle well. Metering block with tube idle well is identified by two aluminum spots and four lead balls on top.

Idle-feed restriction may also be a brass tube in top of idle well (1), or a brass restriction in cross channel between idle well and idle down leg (2).

Idle restrictions may be pressed into the top of the idle well. Brass restriction has been placed on a drill inserted into the actual restriction.

tuning problems. Fortunately, solutions are available. If your racing requires a wild "bumpstick," you should not mind the extra effort required. You want to ensure that the engine will idle at a reasonable speed and not load the plugs when the car is run slowly—as on warm-up laps, caution laps or running back to the pits after a drag run. This work also makes the engine more controllable in approaching the staging lights at the drags. This effort is worth the time it takes.

Pre-Installation Checks–Follow the normal pre-installation procedures of checking the accelerator-pump setting and making sure the pump cover and bowl screws are tight. Look at the underside of the carburetor with the throttle lever held against the curb-idle stop (not against a fast-idle cam). Note the position of the primary throttle plates in relation to the transfer slots or holes. This relationship was established by the factory engineers to give the best off-idle performance. Reasons for these slots/holes are described in the idle system explanation, pages 30-33.

Check the throttle-plate-to-throttle-bore clearance with a feeler gauge or strips of paper as you hold the throttle lever against the curb-idle stop. Note this clearance in your tuning notebook. Record everything as you proceed, regardless of your wonderful memory.

Install Carburetor–Install the carburetor on the engine and start the engine. If you have to increase the idle-speed setting to keep the engine running, note how many turns or fractions of a turn open the throttle to this point. Adjust the idle-mixture screws for the best idle. If the mixture screws don't have any effect on the idle quality, note that fact.

Use a responsive (not highly damped) vacuum gauge to measure manifold vacuum at idle. If manifold vacuum occasionally drops to a value lower than required to open the power valve, a power valve that will remain closed at idle must be installed before proceeding with the next changes.

Drilling Throttle Plates–Take off the carburetor. Turn it over and note where the throttle plates are in relation to the transfer slot. If you can see more than 0.040-in. of the transfer slot between the throttle plate and the base of the carburetor, drill a hole in each primary throttle plate on the same side as the transfer slot.

Start with a 1/16-in. drill on your first attempt and then work up in 1/32-in. steps. Before reinstalling the carburetor, reset the idle to provide the same throttle-plate-to-bore clearance you measured in your pre-installation check.

If holes already exist in the throttle plates, enlarge these holes. Some Holleys are equipped with such holes. In the case of a 400- to 450-cid engine, the hole size required often works out to be 0.125–0.140 in. Smaller engines require smaller holes.

Start the engine. If the engine idles at the desired speed, the holes are the correct size. Too slow an idle means the holes need to be larger; too fast means smaller holes. If holes have to be plugged and redrilled, solder or close the holes with solder, Devcon "F" Aluminum or other epoxy-based material.

It's easier to take the carburetor off to increase a hole size in the throttle plate/s than it is to solder or epoxy the hole closed and drill another one.

When you have the holes correctly sized (which may require using number drills to get the idle where you want it), note the size. Where there are two throttle plates on the primary side, both plates should have the same size hole.

Drilling Idle Feeds—Do the idle-mixture screws provide some control of idle quality? Do they cause the engine to run rough as the idle needles are opened? If not, open up the idle-feed restriction approximately 0.002 in. at a time until some control is achieved. Correct control is indicated when the engine runs as smoothly as possible and turning the idle-mixture screw either way causes rpm to drop and roughens the idle as the mixture is leaned or richened.

Idle feeds are *very small* holes. You'll need a wire-drill set. Proceed in very small increments. Even a 0.002-in. increase in idle-feed restriction of 0.028 in. is an area (and hence, flow) increase of 13%. Wire drills are not used in a power drill. Hold a wire drill in a pin vise and turn it with your fingers. Pin vises are available where you buy wire drills, namely at precision tool supply houses, model shops, or through the *Sears Precision Tool Catalog.*

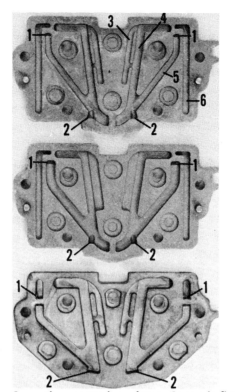 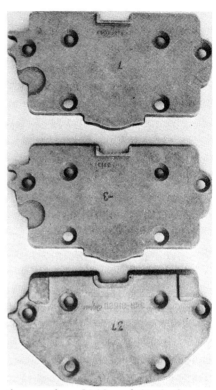

Secondary metering plates. At top is first design used on 3160 and 4160. Center version supplied on some 4160s for Chevrolet had larger well capacity. Aluminum plate (instead of zinc) at bottom is for Chrysler 4160s. Numbers identify features: (1) idle-feed restriction, (2) main restriction, (3) main air well, (4) main well, (5) idle well and (6) idle down leg.

Transfer Fuel Mixture—An adequate accelerator pump setup usually eliminates any need to work on the transfer fuel mixture. This is controlled by the idle-fuel-feed restriction (IFR). The mixture can be checked by opening the throttle with the idle screw until the main system just begins to start. Then back off the screw until it just stops. If richening or leaning with the idle-mixture screw causes rpm drop-off, the mixture is correct. On the other hand, if leaning it causes the rpm to increase, the mixture is too rich and vice versa.

Even if the carburetor is *really too big* for the engine, try it before making any changes to the idle system. If idle and off-idle performance turn out to be unacceptable, increase the idle-feed restriction slightly.

Make changes in very small increments. It is all too easy to drill out certain items of the metering block so there is no easy way to get back to the starting point. Sometimes this requires a new metering block.

If you are in a time bind and have to go racing before you can make the recommended sequence of tests and adjustments, then a quick fix may help. However, it is only part of the cure for this situation in which the throttle is too far open at idle.

Refer back to page 128 where we discussed pump delivery vs. cam type. Pump cam lift is all in by about 20° throttle opening on most of the cams. Thus a wide throttle opening can use up 40 to 50% of the pump shot. On Holley two- and four-barrel carburetors that use a cam for pump actuation, the cam can be rotated to its second positon to regain part of the pump-shot capacity. Only the primary cam is rotated. Even if you use the cam-relocation quick-fix you'll still have to take the time to

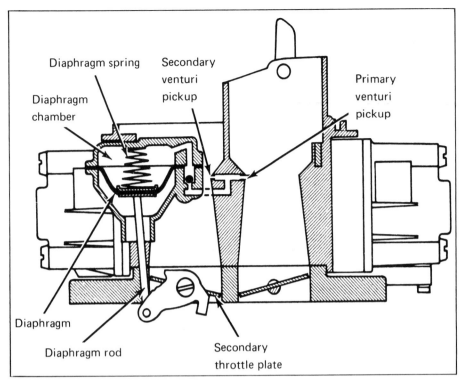

Vacuum secondary schematic.

Diaphragm spring

Secondary venturi pickup

Diaphragm chamber

Primary venturi pickup

Diaphragm

Diaphragm rod

Secondary throttle plate

replace the power valve with the correct one, see page 129.

TUNING VACUUM SECONDARIES

Vacuum-operated secondaries will open at different full-throttle rpm, depending on the spring installed behind the operating diaphragm. A larger engine opens the secondaries sooner than a smaller displacement engine because the larger engine generates higher airflow at the same rpm.

Opening point and rate of opening for secondary throttles can be tailored to the engine and application by diaphragm spring choice. This is the only item to consider changing when working with vacuum-actuated secondaries. This spring counterbalances the vacuum signal from the venturis.

Inconsistent idling can be caused by using a clipped spring or a nonstandard spring behind the diaphragm.

Always start your tuning efforts with the standard spring and then use springs from Kit 20-13 to accomplish

any changes. The yellow spring allows secondaries to open quickest. This spring has the lowest load at a given height. Carburetors are usually supplied with a green, purple or red spring.

If you have the triple two-barrel setup with vacuum-operated end carburetors, you'll have yellow springs in a Chrysler setup; brown springs in a Chevrolet. Play with different springs if you like.

Vacuum Secondary "Tricks"—A number of "tricks" have been touted as important modifications for vacuum-secondary Holleys. Most are almost, if not completely, useless!

The first action many mechanics do is take out the check ball. It's easy to do, even if they don't understand what happens. A bleed groove in the ball seat allows the signal from the venturis to build at a controlled rate so the secondaries don't open suddenly. When the throttles are closed, the ball unseats to bleed the vacuum signal from the diaphragm immediately, thereby allowing the secondaries to close quickly. Closing the throttles

quickly is more important than allowing them to open suddenly.

Without the ball, you can "feel" the secondaries open. Many mistakenly interpret what they feel as an increase in acceleration. Surprise! What they are really feeling is a bog or sag in the acceleration curve.

The recommendation most often heard, and that has destroyed the driveability of more carburetor/engine combinations than any other, is changing diaphragm carburetors to mechanical secondary operation.

A screw is placed in the secondary lever and the diaphragm or spring is removed. Secondary operation is mechanically controlled by the primary throttle shaft.

No pump is available to cover up the hole or bog created when throttles are opened suddenly without regard to engine rpm. This is especially true when the secondaries are operated simultaneously with the primaries (1:1 throttle action). If you want mechanical secondaries for a racing application, then get a carburetor with two accelerator pumps (double-pumper).

Let the engine open the secondaries. Change the opening point by swapping springs if you like. But keep the original spring so you can reinstall it if you don't find improvement through using different ones. Watch the acceleration times closely so you are not misled by the feel in the "seat of your pants."

So you won't spend a lot of time fighting a problem that has been overlooked by many—check secondary-throttle operation by hand as you hold the primary throttle wide open. The throttle shaft should move easily against the resistance of the diaphragm spring and close easily against the stop. Binding can be caused by deposits on the shaft if the secondaries have not been used. And, it is quite common for the shaft to bind due to incorrect and uneven tightening of the carburetor attach-

Quick-Change Vacuum-Secondary Spring Kit 20-59 allows removing diaphragm secondary housing cover with only two screws. Diaphragm springs can be changed quickly. Seven secondary springs with different tension are offered in a Secondary Spring Kit 20-13.

ment nuts/capscrews. This problem arises most often when a soft gasket or a gasket pack has been used.

OFF-ROAD TIPS

Two basic problems are involved: (1) angularity and (2) vibration.

Angularity–Visualize the problem as you tip the carburetor. Note how changing fuel level changes its relationship to the main jets, discharge nozzle and fuel-bowl vents. Note how the angle affects the height of the discharge nozzle in relation to the fuel level.

At some angles, a 40° tilt may cause fuel to drip or spillover from the discharge nozzle. There is a fix: Drop the fuel level approximately 1/16 in.

Problems caused by lowering the fuel level are: Power valve and main jets are uncovered sooner and nozzle "lag" may create a noticeable off-idle bog. Part of this lag can be offset with increased pump. Either open the discharge nozzle/s or "shooters," or use a larger accelerator pump— or both.

Some lag or "bog" is almost inevitable if the carburetor is too large for the engine and rpm being used. Minimize this problem by using a car-

buretor no larger than absolutely required for peak rpm expected in actual driving. Large carburetors are seldom needed for off-road use.

Vibration–The float takes a real pounding in off-road applications. The bumper spring under the float must be selected so it will assist in damping the float's wild gyrations as it vibrates. The spring should be strong enough so it will just allow the float to drop of its own weight.

In some instances this could require using two bumper springs wound together, or a spring from another carburetor. Two springs are offered for side-inlet fuel bowls (4150/60, 4165, 2300). 38R-757 is the standard one with a plain-steel finish.

38R-803 is stronger and has a blue finish.

Vent or Pitot Tubes–Extend these upward as high as possible. You don't want fuel sloshing out of the vent tubes under severe braking and bouncing. Vents can be equipped with small hoses all the way into a remotely-mounted air cleaner.

Throttle Linkage–Use a cable drive because it transmits less vibration to the driver's foot. Many stock automobiles are equipped with such arrangements. It is usually easy to obtain a cable throttle linkage with everything made to fit the job at hand.

Hydraulic linkages are popular, but they can overstress the throttle levers. Hydraulic linkage should be installed

---

AIR CLEANER AFFECTS DIAPHRAGM SECONDARIES

A weaker diaphragm spring is required to get the same secondary opening point when the air cleaner is removed—unless the engine has a very low-restriction air cleaner. The air cleaner restricts airflow, providing a higher vacuum signal to operate the diaphragm at lower rpm. This is typical of diaphragm-equipped carburetors on early Fords and Chevrolets with single-snorkle air cleaners. A yellow spring is a good choice when starting tuning. If secondaries open too quickly, then try other springs for slower action until you find the desired opening point.

so full travel of the actuating cylinder provides WOT. Cylinder mounting must be carefully thought out so there is not an overcenter condition during throttle operation.

Air Cleaner—The best possible air cleaner encloses the whole carburetor, with the throttle operated by a cable. Seal the cable housing where it enters the air-cleaner baseplate.

If the air cleaner assembly is very heavy, support it with special brackets. Avoid overloading the single tie-down stud in the center of the carburetor.

COMMON RACING PROBLEMS

Holley is a major sponsor with representation at races across the country. This includes circle track and drag racing. Tech reps talk with thousands of racers at Holley Clinics and at their trailers where they work on carburetors. From these contacts they've compiled the most common problems. While slanted toward the drag racer, some are common to road racing and circle track, too.

Dirt or Rust—Remove the fuel bowl and there it is: Good old ferrous oxide (rust). The cause is a corroded fuel tank.

Rust is the most common foreign material to invade the fuel system. But sometimes plain old dirt finds its way in through sloppy fuel handling. Whatever the contaminant, flooding usually results because something gets between the inlet needle and seat.

Electric fuel pumps are also susceptible. Contaminants interfere with the internal regulator valve.

A temporary fix is a new unrestrictive in-line filter. Eventually a dirty tank must be flushed and a rusty one replaced. Always use an in-line filter as a safeguard even with a new clean tank.

Fuel Pressure—Most common is running out of fuel at the top end or near maximum WOT rpm, where fuel demand is highest. The problem is an inadequate pump, restrictive system or perhaps the pressure was set too low in the first place. You need a pressure gauge to confirm this. Pressure below 3 psi at the top end is marginal.

You also need a pressure gage to set pressure at idle for systems with an adjustable regulator. If pressure is too high, the flooding tendency is greater. Don't even consider trying to run the Holley high-pressure pump without a regulator! The result is 15 psi to the inlet and guaranteed flooding.

Incorrect Float Level—Remember fuel level should be reset after changing the fuel-inlet valve or fuel pressure. Fuel level must be checked on a new carburetor because shipping shocks can alter float settings.

Accelerator Pump—This is usually adjusted incorrectly. The pump-operating lever should contact the pump lever at idle (or at staging rpm throttle opening) and have a little travel left at WOT. When you've done this you've got a full stroke.

Using Non-Holley Parts—Most common are gaskets, jets and inlet valves. Some are good, but most are not up to Holley standards. Some gaskets are made of less-expensive material and probably not tested with racing fuels. Not all non-Holley jets are flow-checked. Many so-called "high-flow" inlet valves actually flow less fuel than Holley items. Use Holley parts in Holley carburetors.

Removing Power Valves—The only reason for removing the power valve is where the induction system is so restrictive that manifold vacuum at WOT gets above the rating of the power valve. In this case the

valve will close and lean the mixture. The correct solution is to use a higher rated valve. If you remove the power valve you must jet up 6–8 jet sizes to compensate. Then the part-throttle mixture will be 6–8 jet sizes too rich.

Incorrect Jetting—Sometimes this occurs because the power valve has been removed and the basic calibration has been forgotten. Once you've become confused with jet changes for whatever reason, go back to the original jetting with the correct power valve/s and make changes in small increments from that reference point.

Altering Secondary Actuation—Don't open diaphragm secondaries mechanically. Buy a double-pumper if you need mechanical secondaries. Or change the secondary diaphragm spring to get the opening characteristics you want. Use a stopwatch or timer to evaluate changes.

Dirt-Clogged Air Bleeds—When this happens the metering of the idle and/or main systems is altered. Try squirting them with a spray-on solvent. If they are clogged badly, remove the bowls and metering blocks, spray with a solvent, and then blow out the passages.

Carburetor Worn-Out—It's amazing how long some carburetors stay around going from car-to-car and hand-to-hand. After a few years they are in sad shape. You have to make up your mind whether to replace or repair. If the carburetor is the size and type that is just right for your car and the castings and shafts are sound, you might want to repair it. Read the section for your carburetor in Chapter 10, *Adjustment & Repair*. Remember to get a Holley Renew kit before you start.

Alcohol Modifications 9

ALCOHOL STRUCTURE

Alcohols are a partial oxidation product of petroleum and are not found in basic crude oil. To get technical, from a chemistry point of view, the alcohol compound contains a basic paraffin hydrocarbon radical with a hydroxyl radical attached.

As an example, the formula for methane or natural gas is CH_4. This means there is one carbon molecule and four of hydrogen. The formula for methyl alcohol or methanol is CH_4O. You can see that we've added oxygen. That's what we mean by the partial oxidation of a paraffin hydrocarbon.

Ethane, also a gas, has a formula C_2H_6. Add oxygen and you get C_2H_6O or ethyl alcohol, known also as *ethanol*.

We could go on and on. Add oxygen to any of the paraffins and you get the corresponding alcohol. The last time we counted there were well over 20. But, for our purposes, let's limit our discussion to the first two: methanol and ethanol.

Ethanol–The chemistry for this type is explained above, now let's get to the nitty-gritty. Ethanol or *grain alcohol* is so named because it is derived from the fermentation or distillation of fruits and grains. The alcohol in beer, wine and all types of distilled spirits is ethanol.

But ethanol can be used as a fuel. We won't write the combustion formula, but if ethanol is burned perfectly or stoichiometrically, the air/fuel ratio is 9.0:1. This means 9 pounds of air are consumed for every pound of ethanol.

Metering block for Model 4150 alcohol carb has 1/8-in. hole above normal main jets. These holes connect into main well to supply additional fuel flow required when running alcohol. Jet changes are still made with the main jets. Fuel inlet valve has steel inlet needle, 0.130-in. inlet.

The stoichiometric mixture for gasoline is 14.7 pounds of air per pound of fuel. We must put in 1.63 the amount or 63% more ethanol than gasoline per pound of air. Consequently, fuel metering orifices will need to be larger, but only about 56% larger in cross-sectional area because ethanol has a slightly higher *specific gravity* than gasoline.

The heating value of ethanol is about 12,800 Btu per pound. (*Btu* stands for British thermal unit, the amount of heat required to increase the temperature of one pound of water one degree Fahrenheit.) Which means that 12,800 Btu of energy are released

when we burn one pound of fuel. When the engine burns one pound of air, 1,422 Btu are released.

Now examine gasoline's values. It has a heating value of 20,700 Btu per pound. It releases 1,408 Btu when the engine burns one pound of air. In this case, it is a wash. We gain no increase in energy when burning ethanol.

Ethanol's octane rating is 106, while normal unleaded pump gasoline runs about 87, a definite plus for ethanol. The octane numbers we're using in this chapter are *research numbers*. Octane is also rated by *motor numbers*, which run about 10% lower. The advertised numbers have traditionally been the higher research numbers, of course. The number on the pump is an average of the two. Are you totally confused? Anyway, we'll use the higher research numbers in all cases so we're comparing like values.

During the past two crude-oil shortages there was considerable interest in ethanol as a motor fuel. Our country has great capacity beyond our current output to produce grains of all types and so produce ethanol. The process is expensive and would only be practical if the price of crude increased many times over its value at the time of this writing.

Ethanol is excellent as a gasoline extender, that is, added to gasoline to conserve its usage. This product is called *gasohol* and is very successful in extending octane rating as well as gasoline usage. There are usually no adverse effects to the fuel system so long as the alcohol content is held to 10% or less. Gasohol has been used quite successfully since the mid-'70s.

But there is one major problem when using alcohol as a blend with gasoline. The alcohol in the mixture absorbs water, either vapor or liquid, from its surroundings. Unfortunately, after the water reaches a certain concentration, about 1/2 of 1%, alcohol and water separate from the mixture with the gasoline. They become stratified, with the alcohol-and-water mixture forming the lower layer, because it is heavier.

You can picture what happens. The heavier water-and-alcohol mixture goes to the bottom of the carburetor fuel bowl. This causes a lean condition and the vehicle just won't run. This is not a hypothetical case. It actually happens, given the right concentration of water.

Methanol–Methyl alcohol or *methanol* is produced from natural gas, coal or practically any form of biomass, including forest products, agricultural or municipal waste. Gas and coal are in abundant domestic supply. And other feed stocks are renewable.

With perfect burning conditions, as described earlier, the air/fuel ratio for methanol is 6.4:1. Because gasoline requires 14.7:1, we need 2.30 times the amount of methanol or 130% more than gasoline for every pound of air.

The heating value of methanol is 9,800 Btu per pound. This means that 1,530 Btu are released when the engine burns a pound of air. Remember, we said earlier that only 1,408 Btu were released by gasoline when the engine burned one pound of air. This amounts to almost a 9% gain in energy at equal airflow rates, a definite advantage. Of course, we have to pay the price of adding 130% more fuel. Again, because methanol is a little heavier, metering orifices need only be 120% larger in cross-sectional area.

Methanol's octane rating is about 106, the same as ethanol. Like ethanol, it has been used as a gasoline extender in concentrations as high as 10%. We have to be careful with concentrations much higher than this, because corrosion and attacking synthetic materials such as diaphragms and accelerator pump cups occurs.

Also, remember that the carburetor's orifices need to be much larger, so we can expect some degradation of vehicle driveability when using unmodified carburetors with concentrations of methanol as low as 10%. The stratification problem caused by absorption of water, as described above, is also true of methanol.

So what's methanol's primary advantage? It can be produced at about 1/3 the cost of ethanol.

METHANOL AS AN ALTERNATE FUEL

Methanol is probably the best bet as an alternate fuel if a crude shortage and high costs ever appear again. This is likely, given that there is a finite amount of crude available in the world. You'll hear all kinds of opinions as to what that number is, but it is a finite quantity.

All of the major auto companies and a lot of others have programs evaluating the feasibility of running with "neat" or 100% methanol.

Advantages–Besides being the best candidate economically for mass production, methanol has some other advantages because of the characteristics stated above.

Because more energy can be attained per pound of air, greater power output can be attained with the same engine. And the increased octane opens several performance possibilities. Ignition timing can be advanced and compression ratios increased. This puts more "tune" back in the engine and increases efficiency.

Disadvantages–Methanol is toxic: avoid breathing the fumes and contact with your skin (use rubber gloves when handling it). Of course, you can't drink it either. It's also highly corrosive, especially in the presence of aluminum and brass. More expensive materials, such as stainless steel, have to be used in the fuel-supply and delivery systems. In addition, castings must be nickel-plated. All this adds significantly to production costs.

Most of the lower-cost elastomers and gaskets designed to be compatible with today's gasolines are attacked by methanol. They will have to be replaced with more exotic materials such as fluorocarbons, again adding cost.

Fuel-flow capacity must be increased by 130%. This means larger metering orifices in carburetors and higher capacity injectors in fuel injection systems. Fuel pumps must have higher capacity. Fuel-supply lines must be larger and fuel filters less restrictive. To attain the same cruising range, fuel tanks must be larger, which is a real problem with today's downsized vehicles.

There are some operating problems at both cold and hot temperatures. Because the boiling point of methanol is 149F (65C), no vapor is available at low temperatures, making cold starting very difficult. To over-

138

come this problem, the fuel is "spiked" with 10% isopentane, which has a much lower boiling point.

Another problem occurs at the other end of the thermometer–hot starting. Gasoline is a heterogeneous mixture of many different hydrocarbons, each having a unique boiling point. So, during shut-off, after running the engine in high ambient temperatures, the fuel in the tank and carburetor bowl boils or vaporizes gradually as temperature rises.

Not so with methanol, which is a single homogeneous compound. There is virtually no vaporization until the fuel bowl reaches 149F (65C). At this point, *all* of the fuel will boil, causing percolation and vapor lock. A fuel-bowl temperature of 149F is easily reached in southern climates in the summer. These problems can be overcome by increased venting, vapor-return lines and an in-tank electric fuel pump.

Methanol has poorer lubricative properties than gasoline, therefore an additive must be used to prevent excessive wear to valve seats and guides.

In summary, *methanol is not a practical replacement for gasoline* in existing production vehicles. It is, however, a viable fuel when used with a fuel supply, induction and exhaust system specifically designed for methanol. For methanol to work as an alternative to gasoline, commercial fuel handing systems have to be greatly changed. Vehicle costs increase because more expensive materials are used and special vapor-handling mechanisms are employed. Nevertheless, it is a fuel that can be inexpensively manufactured if the supply of crude oil should dry up or become much more expensive.

RACING WITH METHANOL

Because of its low manufacturing cost compared with the other alcohols, methanol is the only alcohol practical

as a racing fuel. So, we will limit our discussion to it.

We stated before that methanol has greater output per pound of air and the octane rating is higher. Add to that the fact that methanol burns cooler and cleaner and we have a very popular racing fuel. A natural. In fact, the fuel has been widely used in circle-track racing for quite some time, especially with fuel injection. Currently, the fuel also sees use in drag racing and specialty events such as tractor pulls.

During the fuel crisis of 1981, various racing sanctioning bodies became concerned about the adverse publicity surrounding the use of gasoline in motor sports. Remember, this was when the average driver had to stand in long lines to get fuel, if he could get it at all. Never mind that the fuel used in racing was an insignificant amount compared with the total usage. Plus, some racers were having trouble obtaining the fuels they desired and costs were rocketing.

The opportunity existed to show the concern that the racing industry had for the problem and to prove that

the use of alcohol in an engine was practical. A chance to take a leadership position was at hand.

Holley was encouraged by these sanctioning bodies to supply racing carburetors designed and calibrated to run on methanol. It responded with several carburetors. In the Competition Carburetor series, the Model 2300 two barrel, List number 0-9647, flows 500 cfm; There are three Model 4150 four barrels: List 0-9645, flows 750 cfm; and the 0-9646 has 850 cfm capacity. In the Pro Series carburetors, Holley's list 0-80498 flows 950 cfm.

Carburetor Modifications–These are based on existing models with modifications to adapt them to methanol. The main wells, metering-body cross channels and booster IDs were opened up as much as possible while staying within the confines of their respective castings. The Pro Series carburetors use a different main-body casting that allows the use of larger passages throughout.

A supplemental drilling of 0.125-in. diameter was added from the fuel bowl to each main well. This drilling

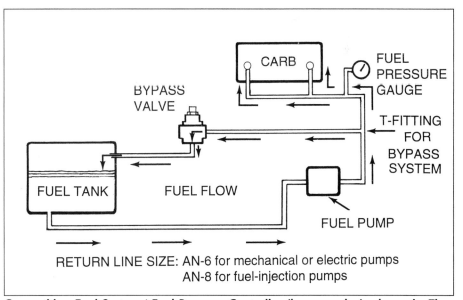

Competition Fuel Systems' Fuel Pressure Controller (bypass valve) schematic. They modify a Holley Hi-Pressure regulator to ensure steady fuel pressure to carburetor. Install valve as shown and ignore "IN" and "OUT" on side of bypass valve. These no longer apply. Connect fuel-pressure gauge to side of fuel distribution T-fitting at the carb. Set the fuel pressure with engine idling. Run in adjustment screw to increase pressure and back out to decrease. Then reset float levels in carb.

Holley 800-cfm carburetor modified for alcohol use by Competition Fuel Systems of Tucson, Arizona. Features 910 cfm airflow, removable air bleeds and idle adjustments for each barrel. Special throttle linkage allows adjusting secondary action to tune throttle response off the corners. Bottom view photo shows holes in each throttle plate to get correct relation of throttle to idle system for use with racing camshafts, as described on page 131.

is just above each main jet. The purpose is to introduce more fuel into the main system without requiring larger main jets. The normal Holley main jets couldn't be made larger.

Needles and seats are steel with an opening of 0.130 in., the largest possible. Holley recommends idle fuel pressure of 9 psi to ensure maximum possible fuel flow into the bowl. Any higher pressure would invite fuel-handling problems. Brass floats are used because Nitrophyl will not hold up in methanol. New hollow plastic floats were added in 1987.

All accelerator pumps are the 50cc variety with pump shooters around 0.042-in.

No special plating was added to the castings and the pump-diaphragm material remained rubber. However, in 1994, the Pro Series carburetors were introduced with new GFLT diaphragms that are compatible with alcohol-based fuels. It is recommended that racers tear the carburetor down after each meet to clean it. Pay special attention to the diaphragms and gaskets and change them as needed. Holley recommends that after each use with alcohol, the fuel system be purged with gasoline.

Holley recommends using a 12-705 or 12-706 Pro Series Volumax™ electric fuel pump with external regulator with a 12-704 or 12-707 external fuel-pressure regulator. These items are designed to allow use with alcohol-based fuels. You could also use the 12-802 electric fuel pump with the external regulator. Two reasons here: the pump has few rubber parts and has the capability of providing the 9 psi needed. Clean the pump and install a new repair kit at frequent intervals.

However, there is a problem with using the 12-802 pump with alcohol. The rubber gasket between the body and cover plate can swell. Use this gasket as a template to make one from a fluoro-silicone gasket material.

With multiple carburetion use one

pump and one regulator for each carburetor. Use 1/2-in. steel fuel lines, *never* rubber hose.

In summary, methanol is an excellent racing fuel. Anyone using it should be aware of its toxicity and the problems when using it, and be prepared to deal with them.

All fuel-metering orifices must be larger and fuel-supply capacity needs to be much greater than gasoline. Corrosion and incompatibility with some rubbers and plastics must be considered and adapted to.

If a pushrod pump is required, use one of Holley's Pro Series mechanical fuel pumps. Or, if there is no rule limiting fuel-pump type, use a Hilborn 150-A fuel-injection pump as described later in this chapter.

SHORT-TRACK RACING WITH METHANOL

Norm Schenck of Competition Fuel Systems, 3525 S. Palo Verde, Tucson, AZ 85713, has years of experience supplying and modifying Holley methanol carburetors for racers. The remaining text in this chapter was written by Norm. It's based on his knowledge gained from research and development on racing methanol carburetors and from working directly with racers.

"Because methanol carburetion applies itself best to oval short-track racing, and because that is where the majority of methanol carbs are being used, most of my comments are about running methanol in short-track racing engines."

Methanol Improvements–"When we compare gasoline to methyl alcohol (methanol), the most important factors to consider are improved torque and horsepower (hp), and cooler engine temperatures.

"When methanol is run in American wedge-chamber engines, we usually see a mid-range torque increase of 8–12% over gasoline, depending on compression ratio, valve timing, com-

bustion chamber design and many other factors. Even with some of these being far from optimum, there will be, in most cases, a significant torque increase with methanol. At the majority of short tracks, 'off-the corner' torque is the most important factor of the engine's output, and this is the major reason for the popularity of methanol in this type of racing.

"Top-end hp increases with methanol range from moderate losses to increases around 10%. The reason for this wide range is that high-rpm combustion efficiency varies widely from one engine to another. With methanol, good combustion efficiency is more difficult to achieve because methanol is a slower burning fuel. Compared with gasoline, methanol will usually require 3–5° more ignition advance and a higher-energy spark to get combustion properly started and progressing.

"To allow the flame travel to progress across and around the combustion-chamber volume, combustion-chamber and piston-dome design must not impede the movement of the flame front any more than is necessary to achieve the desired static compression ratio. The higher the compression ratio, especially over 13:1, the more critical chamber and dome shapes become if good high-rpm flame travel is to be maintained.

"Generally, the faster the flame travel procedes through the combustion volume, the more complete the combustion will be, thus making more cylinder pressure (during the power stroke) available to produce power at the flywheel.

"Higher compression ratios will nearly always increase mid-range torque because there is adequate *time* available for flame travel. But improvements in high-rpm hp *may* be *little* or *none* if high-rpm flame travel is slowed by a larger piston dome used to increase the compression ratio.

"The characteristics of methanol can be used to help cool the engine.

Methanol's richer air/fuel ratio cools in two ways. First, the increased volume of fuel absorbs more heat energy from the intake system in the process of vaporizing, and second, the combustion temperatures with methanol are generally 100–150F (38–65C) lower than with gasoline.

"The price to get the power and cooling benefits from methanol is paid in the components (carb, fuel pump, fuel lines and so forth) necessary to deliver the larger volume of fuel to the engine. Compared with gasoline, methanol fuel flow will be from 2–2.5 times higher, depending on the combustion efficiency of the engine and cooling efficiency of the cooling system in the race car.

"A convenient term to describe fuel-flow requirements is Brake Specific Fuel Consumption (BSFC). This is the fuel flow (in lb./hr.) at a particular rpm divided by the observed hp at that rpm. It is stated as "lb. per hp-hr." The BSFC of short-track engines in track tune is usually around 1.0–1.2, but has been seen as good as 0.88 and as bad as 1.35. An engine with good combustion efficiency and cooled by an efficient cooling system can have a BSFC at the lower end of this range without having overheating or detonation problems."

"An efficient engine will usually make its best mid-range torque with an air/fuel ratio a little on the rich side. Best top-end power usually occurs with a slightly leaner mixture. To give the engine of the best torque and best horsepower on the same dyno "pull," the carb must gradually lean the air/fuel ratio as rpm increases. This sort of air/fuel-ratio curve will generate a fairly "flat" BSFC curve. An example of a "flat" BSFC curve would be a BSFC of 1.0 at peak torque and 1.05 at peak hp.

"BSFC is an indication of the engine's combustion efficiency. It is not an absolute indicator of air/fuel ratio. For good comparisons, both

BSFC and air/fuel ratio should be monitored simultaneously to get an idea of whether BSFC changes due to a change in air/fuel ratio or because of changes in engine efficiency. Air/fuel ratio is only one of several factors that affect BSFC.

"If combustion or cooling system efficiency falls off, more fuel will be needed to maintain the engine at a reasonable temperature. This increases fuel consumption and lowers torque and hp, making the BSFC numbers rise quickly into the inefficient end of their range—over 1.15. This demonstrates that the myth about not needing a good cooling system when running methanol is false, particularly when you are running a track that will accept as much torque and hp as your engine can make."

ENGINE MODIFICATIONS

Norm continues: "There are other engine modifications, in addition to increased compression ratio and total ignition advance, that are needed to achieve all of the potential response and power improvements of methanol over gasoline.

Intake Manifold–"One of the most important considerations is the size and shape of intake manifold and cylinder head intake passages. Because methanol/air mixtures are so rich with fuel, some potential airflow of the engine is being 'crowded out' by the fuel.

"To compensate for this requires an increase in the volume of intake plenum and ports. But this port volume increase must be approached carefully so that mixture velocity doesn't drop low enough to affect the volumetric efficiency (VE) of the intake stroke, or to allow some of fuel to 'drop out' of the mixture flow and 'puddle' in the bottom of the plenum.

"Another requirement to keep the fuel in the airflow is to keep the number of directional changes that the fuel/air mixture has to make on its

way to the cylinders to a minimum. And make each of those changes as gentle as possible (i.e., largest radius curvatures possible). For this reason, most short-track engines will work better with a single-plane intake manifold, because the mixture flow has fewer and gentler turns compared with other manifolds.

"The temperature of the intake system needs to be higher with methanol for two reasons. First, a higher temperature encourages even mixture distribution, minimizes fuel 'drop out' problems (at lower engine speeds) and maximizes combustion efficiency. It is necessary to vaporize as much of the fuel (in the fuel/air mixture flowing through the intake system) as possible before it gets to the cylinders.

"A methanol/air mixture (at its proper fuel/air ratio) requires about 9 times more heat energy (compared with a gasoline/air mixture) to vaporize the fuel. Most of that heat energy must be provided by the intake system. The cooler the intake air temperature above the carburetor, the warmer the intake system needs to be to compensate.

"In weather where the air temperature is below 50F (10C), heating the area under the intake manifold plenum with engine coolant improves engine performance through the entire rpm range, particularly low speed driveability, and mid-range response and torque.

"The second reason for making the engine's heat available to the fuel/air mixture is to take some of the cooling load off of the cooling system by absorbing some of the heat generated by the engine to vaporize the fuel in the intake system.

"The typical short-track V8 engine running methanol will run best when the engine coolant temperature is between 190–210F (88–99C). As stated before, the racer will be hp ahead by making the cooling system as efficient as possible. And, in many cases, you should use a thermostat to *regu-*

late engine temperature–rather than using excess fuel to absorb the heat the otherwise inefficient cooling system can't handle."

Ignition Advance–"Modifications to engine ignition advance must be done to compensate for the slow-burning characteristics of methanol, especially at part-throttle conditions with engine speeds below 3000 rpm. To improve low-speed driveability and response with methanol, I *decrease* the amount of mechanical (centrifugal) advance in the distributor so that I can *increase* the amount of initial advance at the engine's idle speed.

"For example, a typical *high-compression* short-track small-block Chevrolet V8 engine will require at *least* 20° initial advance at idle speed (to idle and respond properly), and around 36°–38° total advance (all in by 3000 rpm at the latest). Some engines respond better to completely *locked* timing (no mechanical advance at all) to get a smooth steady idle, and sharp low-speed response and driveability.

"Generally, the smaller the engine displacement and/or the longer the intake duration of the camshaft, the more initial timing needed to get good idle and low-speed characteristics."

Connecting Rods–"To extend the torque improvements of methanol, many engine builders are using longer-than-stock connecting rods to increase the rod length/stroke ratio of the engine. By increasing this ratio, engine torque above 3000 rpm can be improved because higher cylinder pressures are generated during the power stroke. When this 'rod ratio' is properly *matched* to the engine's application and to the other components (heads, camshaft), torque improvement will be achieved."

Camshaft–"Camshaft selection for a methanol engine, especially one with a higher rod ratio and larger intake port volumes, is one of the most critical component selections. The cam

142

must work with everything else in the engine to make the intake and exhaust cycles of the engine as efficient as possible through the entire rpm range the engine will see on the track.

"When rod ratios are higher and port volumes are larger, it is much easier to 'overcam' the engine. In particular, a little too much intake duration can really limit mid-range torque. This is because increased duration, combined with larger ports and longer rods, can allow the fuel/air mixture velocity in the intake ports to drop below a certain level. Below this level the VE of the intake cycles of the engine drops quickly. When the proper intake profile is used, that torque comes back, but check too that there is no loss of top-end hp."

CARBURETOR MODIFICATIONS

"When all of the engine-component selections and modifications are done, all that information and some 'at-the-track' data should be used to select and modify the proper carburetor for the engine. This is just as important as camshaft selection when trying to get all of the engine components to work together to produce as much power as possible.

"To enable the carburetor to meter the proper amount of fuel needed to produce that power, all fuel-delivery components (pump/s, pressure controller/s, fuel lines, fittings, and so forth) must be sized for adequate fuel-flow capacity under all racing conditions."

Metering Orifices–"The 'methanol modifications' I do to make a better running Holley methanol carburetor are all designed to allow it to meter the proper amount of fuel under all conditions. Because this amount is 2–2.5 times higher than when metering gasoline, all of the diameters of the metering orifices (jets) and passageways must be increased by about 1.5 times over the proper diameters

for gasoline. In many cases, the new passageway diameters required in the metering blocks are very close to the physical limits of the block.

"Special techniques have been developed to increase the flow capacity of these passages, especially on the larger carbs, i.e., 850 cfm 4150s and 1050 cfm Dominators. The most important passage is the one in each venturi booster that carries fuel out to the middle of the venturi. Some boosters have the capability of tolerating a larger passage; others don't."

Booster Signal–"Once the passage has been adequately enlarged without compromising any physical aspects of the booster, then the booster *signal* must be checked to see if it is 'usable' with methanol. This signal is the vacuum draw the booster generates and uses to pull fuel up through the main metering circuits to the venturis (where it is mixed with the incoming air)

"The booster signal curve generated by the venturi boosters through the usable airflow range of a carb is the *heart* of that carb. To meter fuel correctly, the booster signal curve must be tailored to such things as main-circuit passage diameters, needle-and-seat flow, fuel-pressure capabilities, etc.

Needle & Seat–"The inlet needle and seat assemblies for a methanol carburetor should be chosen to handle fuel pressure and the required flow. They must be matched to meet the pressure capability of the fuel pump at the maximum fuel flow required. And they have to handle the maximum allowable amount of fuel level drop (*float drop*) for the particular carb being used.

"Venturi booster signal characteristics, main-circuit passage diameters, and needle-and-seat flow capacity are interdependent, i.e., one can partially compensate for inadequacies in the other. For example, in my experience, the Holley four-barrel methanol carb, 0-9645, tends to run too rich in the mid-range and too lean on the top end because of:

• The factory main jet calibration.
• Inadequate needle-and-seat flow capacity.
• Inadequate (too small) diameter in the venturi booster passage.
• Inadequate booster-signal curve

"Main-jet calibration and needle-and-seat flow are easy to change. But the booster passage is much more difficult to change because of the booster's design. By using a needle-and-seat design with more flow capacity than would normally be needed in a carb this size, we can partially counteract the top-end lean-out tendency caused by the small booster-passage size.

"By changing the main jets and needle/seat assemblies, we can get a fuel-delivery curve that is safer and more competitive, even though it isn't as good as could be achieved if the booster passage were larger."

Fuel Delivery System–"Just as important as having proper passage sizing and flow capacity in the carb is the fuel-flow capacity of the fuel-delivery system: pump, line and pressure regulator. This system must deliver the correct amount of fuel to the carb at the correct pressure. This is the pressure needed to flow the correct amount of fuel through each needle/seat assembly with a reasonable amount of float drop."

Fuel Pump–"The six valve mechanical (pushrod) fuel pump, commonly called the *NASCAR pump*, has fuel flow capacity (at 7 psi fuel pressure) to safely handle about 550 hp with average maximum power BSFC of 1.15–1.20. If maximum power BSFC is lower, the *hp capacity* of the pump can be over 600 hp, but there is no safety factor. That is, no reserve capacity to compensate for *any* problems, such as a slightly clogged fuel filter.

"Engines in this hp range that must run this type of pump by rule, must be designed for maximum combustion efficiency to run safely. The entire fuel system must get daily and weekly maintenance, and the carb must not have any tendencies toward a top-end lean-out, which is the most common reason for engine damage.

"For those classes that don't specifically require a pushrod fuel pump, engines over 550 hp should use a belt-driven fuel-injection pump, i.e., the Hilborn 150-A. It's called the dash-zero pump. EVM, Enderle and Kinsler also make suitable pumps. The extra expense for this system should be viewed as a worthwhile insurance policy to protect the healthy investment already made to get the engine to this power level."

Fuel Line & Filters–"Fuel lines should be at least AN-10 size, including the pickup tube in the fuel tank (cell). Fuel filter/s should be of the highest flow capacity available. And filter elements should be cleaned or replaced at least once a week, particularly in the busiest part of the racing season."

Pressure Regulator–"This should not restrict *fuel flow* to the carb at full power. For this reason, many racers use a bypass valve system that controls fuel pressure by regulating fuel flow back to the fuel tank through a return line. This results in higher full power fuel pressure because there is no longer a flow restriction between the fuel pump and carb–a necessary design to help prevent top-end lean-out."

Fuel System Insulation–"Another very helpful, though less-obvious modification to prevent lean-out and engine heating problems is to insulate the fuel lines, pump, and filter. By keeping the radiant heat of the engine, exhaust system, and air (from the radiator) away from the fuel in the delivery system, the temperature of fuel getting to the carb will be lower and more constant through the race.

"This lower fuel temperature keeps the specific gravity (density) of the fuel higher, and makes the carb's job of metering the proper amount (lb./hr.) of fuel easier. Without insulation, a cycle of rising fuel temperature and engine-coolant temperature can occur: Both temperatures keep rising during the race until the cooling system or the engine fails."

RACING PROBLEMS

Water In The Fuel–"Many problems racers have with running methanol are caused by the lack of regular fuel-system maintenance. The characteristics of methanol make it harder on all fuel-system components than gasoline. Methanol absorbs water moisture directly from the air, even through the fuel-tank vent and carb-bowl vents.

"The water content in the methanol, if allowed to get high enough, can cause fuel-metering lean-out problems in the carb, flow restrictions in pleated-paper fuel-filter elements, and severe corrosion of the metal parts in the fuel system.

"To prevent these problems, I recommend that the fuel remaining in the car's fuel cell, lines, pump, and carb fuel bowls should be drained out and put into an airtight container to minimize any further water contamination. This should be done as soon as possible after each day's racing is over, even if the car is to be raced the next day.

"A good way to analyze methanol for water contamination is to check the specific gravity of the fuel with a hydrometer. At 60F (16C) fuel temperature (standard checking temperature for fuels), pure methanol has a specific gravity of 0.792. If your methanol (at 60F) has a specific gravity over 0.800, discard it and obtain some new fuel–and check it too."

Adjustment & Repair

ANALYZING CARBURETOR PROBLEMS

There are three reasons to take wrench and screwdriver in hand and lay your carburetor open from choke plate to idle screw.

1. You're curious.

2. You're a racer or performance enthusiast.

3. You've got a driving problem with your vehicle and want to fix it.

If you are a number 1 or 2, skip to the appropriate section on disassembly and repair. If you belong to group 3, this section was written *just* for you.

How can you be sure the problem is with the carburetor, or involves it at all? An experienced carburetor tuner we know claims he's fixed a lot of "carburetor problems" by changing spark plugs and distributor points. There's lot of truth in that statement. Replacing plug wires is another instant "magical" fix.

When analyzing engine malfunctions you have to look at the *total* system. That's what the next few pages are all about. Common problems associated with carburetors are presented with some probable solutions. Look under the *Problem List* for the malady that is occurring. The numbers refer to possible answers in the *Solution List*.

We also list non-carburetor causes for a simple reason: Disassembly and repair of a carburetor is truly time-consuming. It sure is disappointing to go through the whole exercise only to discover later that the problem is in some other component or system. Our objective is to get you to the solution quickly and simply.

CARBURETOR TROUBLESHOOTING — POSSIBLE CAUSE	Stalling	Rough Idle	Flooding	Hot Start	Economy	Hesitation	Acceleration	Surge	Back Fire (cold)	Power	Stalling (cold)
Idle Adjustment	X	X		X	X	X		X	X		X
Idle Needles (Damaged)	X	X									X
Idle Vent Adjustment	X	X		X							X
Fast Idle Adjustment	X										X
Idle Passages (Dirty, Plugged)	X	X				X		X	X		X
Auto. Choke Adjustment										X	X
Choke Diaphragm Adjustment										X	X
Choke Rod Adjustment										X	X
Choke Unloader Adjustment										X	
Metering Jets (Loose, Plugged)					X	X	X	X	X	X	
Power Valve (Loose, Sticking						X		X		X	X
Fuel Inlet Needle & Seat (Loose, Leaking)	X	X	X	X	X	X				X	
Float (Leaking, Rubbing, Wrong Setting)	X	X	X	X	X	X		X		X	
Gaskets (Brittle, Improper Seal)	X	X			X	X		X			
Pump Discharge Holes (Dirty, Plugged)						X	X	X			
Pump Diaphragm (Worn, Cut)						X	X	X			
Pump Ball Checks (Dirty, Sticking)						X	X	X			
Choke Diaphragm Adjustment (Vacuum Leak)										X	X
Choke Valve & Linkage (Dirty, Sticking, Damaged)										X	X
Secondary Carb. Linkage Adj.						X	X		X	X	
Secondary Lockout Adjustment						X			X		X
Throttle Valves (Loose, Damaged, Sticking)	X	X				X		X			X
Venturi Cluster (Dirty, Loose)	X			X	X	X	X	X	X	X	X

Chart specifies carburetor problems only. It's assumed engine is in good mechanical condition and in tune. Many ignition and carburetor problems have same symptoms. Don't assume fault is with carburetor.

PROBLEM LIST

A. Hard starting–cold engine:
   1, 2, 3, 4, 5, 6, 7, 8, 9, 10, 11.
B. Hard starting–warm engine:
   1, 10, 12, 13, 14, 15.
C. Rough idle and stalls:
   4, 7, 8, 16, 17, 18, 19, 20, 21, 22, 33, 53.
D. Deterioration of fuel economy:
   1, 4, 7, 18, 20, 23, 24, 25, 26, 30, 49, 52, 53.
E. Sag or hesitation on light accelerations:
   8, 10, 14, 18, 24, 25, 27, 28, 30, 32.
F. Sag or hesitation on hard accelerations:
   23, 27, 29, 30.
G. Surge under cruising and light loads:
   8, 10, 14, 18, 24, 25, 26, 30, 31, 32.
H. Surge at high speeds and heavy loads:
   23, 25, 26, 31, 34, 35.
I. Misfire or backfire:
   1, 4, 21, 27, 31.
J. Loss of power and top speed:
   1, 14, 18, 20, 23, 31, 34, 35, 36.
K. Fast idling or inconsistent idle return:
   8, 19, 37, 38, 39, 40, 41, 42, 43, 44.
L. Vapor lock—loss of power, stall or surge on acceleration after short engine-off period in high temperatures:
   10, 34, 35.
M. Inoperative secondary system (diaphragm-operated):
   45, 46, 47, 48.
N. Engine bucking or emitting black exhaust smoke:
   1, 23, 49, 50, 51, 52.
O. Poor cold driveaway:
   1, 2, 3, 29, 30, 51.

SOLUTION LIST

1. Binding or sticky choke plate or linkage. This can sometimes be repaired without removing the carburetor. If you have a remote or divorced choke, disconnect the choke rod from the carburetor to isolate the problem. *Never use lubricating or household*

Mass confusion! An excellent example why you absolutely must identify all hoses and connections. Make yourself a sketch for reassembly. Your memory is guaranteed to fail when you try to reconnect all these hoses!

oil to free the choke. Oil gathers dust and grime and stiffens at low temperatures, creating an even worse problem. Commercial solvents for carburetors work well. Penetrating oil works just great.

Sometimes vibration or a few backfires will cause the choke plate to shift on the choke shaft. Repair requires loosening the choke plate screws and realigning the plate in the air horn. If the choke is held on by screws with lock washers this is easy. Most have staked screws to prevent loosening. Loosening any staked screw requires considerable care to avoid twisting off the screwhead.

Use a small file to remove the upset or staked material from the screw and proceed with caution. You must restake the screws after you retighten them. The opposite side of the shaft must be firmly supported so you won't bend the shaft during staking. Disassemble the carburetor to allow staking the screws. With a divorced or remote choke, the linkage or bimetal in the choke pocket in the intake manifold can rub against the side, causing

binding. Move the choke rod up and down to detect this condition.

2. Incorrect choke adjustment. There are two basic adjustments–the bimetal index and qualifying or pull. See next section.

3. Overstressed bimetal. If the choke unit doesn't exert enough force to close the choke plate when the engine is cold and ambient temperature is below 60F (15.5C), this could be your problem. A new bimetal is required. Because replacement bimetals are not always readily available, try resetting the choke in the RICH direction. Most automatic-choke housings have an arrow indicating RICH and LEAN directions.

4. Fouled or old spark plugs. You can easily spot this with an engine analyzer but not many of us have this tool. Pull a plug or two and take a look. If you see a deposit build-up or burned electrodes on one or two, then change the *whole* set. Spark plugs are inexpensive and the benefits of good firing are numerous, especially for economy. If plugs and wires are over a year old, replace the whole set regardless of how good they look.

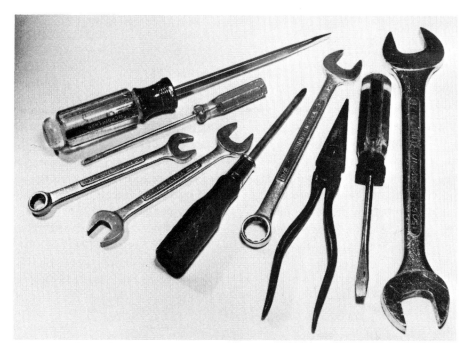

These are the ordinary tools you'll need for working on carburetors. A clutch-head screwdriver is also needed for Holley Model 4160/4180 carburetors. Torx drivers (T20, T25) are needed for 2010, 4010 and 4011 carburetors. Other essential tools include tubing wrenches, Corbin-clamp pliers, feeler gauges and a special fuel-inlet open-end wrench.

WARNING: Be careful when spraying flammable carburetor cleaner on an idling engine–don't let it ignite! Use it sparingly and in a well-ventilated area.

9. Low battery voltage. Slow cranking speed is the giveaway here.

10. Fuel volatility. Oil companies change the *volatility* or *vapor pressure* of their fuels so they are highest in mid-winter and lowest in mid-summer. This allows quick cold starts with little cranking in January and minimizes long cranking to get a hot engine started in July. Volatility is also varied geographically, with higher values in the North and lower ones in the South. An unusually warm winter day always brings a rash of hot-starting problems. Avoid filling up at low-volume gas stations–especially in the spring and fall.

11. Water in the gasoline. Modern gasoline additives make this less of a problem than it once was. Nevertheless, always fill your tank completely to minimize the opportunity for water vapor to condense in the unused volume. It also reduces the likelihood of rust formation in your fuel tank. For the same reasons avoid purchasing fuel at low-volume stations. It is a good idea to add a can of water-absorbing fuel-tank cleaner like Dry Gas or STP Gas Treatment once in a while.

12. Incorrect carburetor fuel level. An excessively high fuel level causes rich die-outs on sudden stops due to fuel spillage. There is also a greater tendency toward *fuel percolation*. It boils over from the bowl into the manifold.

A low fuel level can cause sags when accelerating because it delays the main system start-up. Fuel starvation (leanness) on turns and maneuvers can also result from a low fuel level.

Some Holley carburetors have externally adjustable fuel levels and removable sight plugs. With the engine idling the lock screw should be

5. Wrong distributor point gap or dwell. Detect with a dwell meter or by physical measurement. With the point rubbing block set on one of the cam lobes, the points of most non-electronic distributors should be open about 0.017 in. Check with a feeler gauge or a No. 77 drill. Check the points for pitting or build-up.

Point changes can be accomplished on the engine but it is easier and less risky if the distributor is removed. If you remove the distributor, note the position of rotor and housing and have a timing light handy to re-time the distributor correctly when you reinstall it. You must make sure the distributor drive mechanism fully engages with the oil pump if it drives the pump.

6. Water condensed in the distributor cap. This can happen with warm humid days and cool nights. Remove distributor cap and wipe with a clean dry cloth. Check inside the cap for cracks or carbon tracking while you have it off. Either means a new cap.

7. Poor compression. This condition could result from worn piston rings or leaky intake or exhaust valves. A compression or leak-down test will give a quick answer. Consult a service manual for desired compression values. If one or more cylinders has more than 20% less compression than the others, you've got problems. A leak-down tester is better than an ordinary compression tester because it helps you find the leak.

8. Intake leaks. Look very closely at all rubber vacuum hoses. Make sure they are not split, leaking or disconnected. These hoses become hardened with age. Cracks are often hidden in bends or on the underside, so look carefully. Be sure to check the hoses to the automatic transmission. Manifold-to-head-gasket leaks can be detected by spraying a little carburetor cleaner all around the intersection while the engine is idling. If the engine speeds up or smooths out, you've found the culprit.

loosened and the adjustment nut turned until the fuel level is just at the bottom of the sight-plug hole.

Other carburetors require disassembly for float adjustment. This is treated in more detail for specific carburetors in later sections. The *Holley Illustrated Parts & Specs Manual* gives fuel-level adjustments for most popular Holley carburetors.

13. Leaky fuel-inlet valve. Is the valve worn or is foreign material lodged between the inlet and seat? New inlet valves and seats are included in most repair kits. Fuel levels are noted in the Renew Kit instructions. Always check fuel level when changing fuel-inlet valves or floats.

14. Sticking exhaust-manifold heat-control valve. Most vehicles built before 1969 had heat-control valves. This valve diverts some exhaust gas through an intake-manifold passage to create a "hot spot." A valve stuck in the closed position makes the hot spot run excessively hot. This causes vapor formation and boiling in the carburetor fuel bowl. Hard starting and rough idling result. A valve stuck open will not allow the hot spot to reach high enough temperature. Low and mid-range driveability problems are the symptom.

Free the valve with commercial solvents designed for this purpose. Penetrating oil or WD-40 is especially good. If the valve has been stuck for some time, it may take several applications, a little time and friendly persuasion with light hammer taps.

15. Mechanical bowl-vent adjustment. Bowl vents are often found on pre-'68 carburetors. The vent should be slightly open (approximately 1/16 in.) at idle and engine off and closed at all other conditions.

16. Carburetor icing. Icing can occur when there is high relative humidity and temperatures in the 30–50F (-1C–10C) range. While the engine is warming up, ice forms between the throttle plate and bore. Check the carburetor-air preheat system on 1969 and later vehicles. On earlier models check the exhaust-manifold heat-control valve (see 14, above). A slightly higher idle speed helps minimize icing.

17. Idle-mixture adjustment. Turn the idle-mixture screw in until engine speed drops slightly, then back out the screw about 1/8 turn. This is the *lean-best-idle setting*. For 2- and 4-barrel carburetors, set idle mixture on one side and repeat on the opposite side. Then go back to the original side and repeat the process one more time just to be safe. On 1968 and later vehicles, idle-adjustment range is limited for exhaust-emission purposes.

18. Clogged air bleeds or passages. This usually requires a carburetor teardown just to determine if it really is the problem. Try swabbing and squirting the exposed bleeds with a little Gum-Out, STP Carb Cleaner, other carburetor solvent or isopropyl alcohol. This may open up the blockage and save you a lot of work.

19. Idle-speed adjustment. Vehicles from 1970 and later have recommended idle speed on a tag in the engine compartment. For earlier vehicles, use 600 rpm for manual transmissions and 550 rpm in Drive for automatics.

Be careful when setting the idle on cars with automatic transmissions. Don't stand in front of the car or drape yourself on the fender. Make sure the parking brake is ON and the wheels are blocked. Don't "flick" or "blip" (snap it quickly and then release) the throttle or you could have an out-of-control vehicle.

20. Restricted or dirty air cleaner. Place a 100-watt bulb inside the cleaner element and look at the outside. You can see the dirt pattern. Most times the element can be cleaned with pressurized air. If it looks hopeless, buy a new element. They're inexpensive.

21. Old and/or cracked spark plug wires. With the engine idling on a dark night, open the hood and observe the spark plug wires. If sparks jump from the wires, it's time for a new set. Sometimes the center of carbon-core (TVR) wires can separate. This even happens with new TVR wires. Ignition wires must be removed from the plugs by tugging on the connectors—never by pulling on the cables.

22. Leaky intake or exhaust valves. A compression or leak-down test will usually detect this problem; the exception is a hanging or sticking valve that acts up sporadically.

23. Power valve stuck. A piston-type valve can stick open or closed. Diaphragm-type valves seldom stick, but leaking is possible. A leaky diaphragm causes an abnormally rich condition, evidenced by a black, smoky exhaust and extreme difficulty in restarting. A leaky power valve can be caused on Holley performance carburetors by a backfire. Some carburetors made in 1992 and later may be equipped with backfire protection for the power valve. You can easily add this protection at low cost, see page 55.

24. Distributor vacuum-advance line disconnected or leaky. Hoses become dry and crack with age. Connections should be snug because small leaks have great effects.

25. Incorrect basic distributor setting. Cars from 1970 and later have the recommended setting listed on a tag under the hood. For earlier models, see the service manual.

26. Wrong size main jets. This commonly occurs with racers or enthusiasts, and with used carburetors. Main-jet sizes for most Holley carburetors are listed in the *Holley Illustrated Parts & Specs Manual* and in the *Holley Performance Parts Catalog*.

27. Accelerator-pump failure. Remove the air cleaner and observe the pump nozzles while opening the throttle. A steady stream of fuel should shoot out as you open the throttle. Make sure you install the

Top photo: Staked choke-plate-retaining screw (arrow). Bottom photo: Staked throttle-plate screws. Screws are staked to prevent loosening and falling into the engine. Staking must be removed before taking screws out. Restake with a small punch and hammer on reassembly. Support shaft while staking to prevent bending shaft. Avoid removing staked screws whenever you can. It's seldom necessary for a carburetor rebuild.

Contents of Holley Renew Kit for a Model 4150. Be sure to read the instruction sheet. It contains a lot of tips and adjustment specifications. All necessary parts are provided to rebuild your carburetor to original specifications. Included are gaskets, pump diaphragm/s, inlet valves, power valves, and float gauge. A troubleshooting chart is also included. But you might miss it because it is printed inside the carton. We reproduced this chart on page 145.

fuel-bowl gasket so the pump passages are not blocked.

28. Clogged idle-transfer slots or holes. This condition requires a carburetor teardown to find and remedy.

29. Wrong accelerator-pump adjustment. Specific carburetor sections explain some of the adjustments for your particular carburetor. Renew Kit instructions and the *Holley Illustrated Parts & Spec Manual* give correct settings and describe how to check the settings.

30. Carburetor too large for the application. Usually an enthusiast-caused problem. The *Select & Install Your Carburetor* chapter deals with this subject, page 75.

31. Clogged main jets. Requires internal carburetor inspection. It seldom happens.

32. Hot inlet air system inoperative. This system preheats inlet air by routing it past the exhaust manifold. It is common on 1969 and later vehicles.

33. Clogged PCV valve. This valve is usually installed in a grommet or cap in the valve cover. Inspection is quick and easy. Most can be cleaned with a little kerosene or solvent. You'll know it's clean when you can blow through it in one direction (toward the carburetor) but not the other. Be sure to install it correctly.

34. Clogged fuel-inlet filter or screen. In some cases this can be inspected without removing the carburetor. Some vehicles have an in-line filter between the fuel pump and carburetor. Don't hesitate to change it if you suspect it.

35. Low fuel-pump pressure. A pressure gauge can be installed between the fuel pump and carburetor. Typical idle pressures are 3.5–6 psi. Less than 2.5 psi at top speed is a sure danger signal. *Don't plumb fuel lines to a pressure gauge inside your car!* Mount a fuel-pressure gauge on the cowl outside of the windshield.

36. Throttle plates not reaching wide-open position. All carburetors have a wide-open stop on the throttle lever. Have someone depress the throttle pedal to the floor while you check to see if the throttle lever reaches the wide-open stop. If it doesn't, readjust the throttle linkage or cable. Hold the throttle lever against the stop

Spray-on gasket remover is almost essential to remove stubbornly stuck on Holley gaskets. These are by Permatex and CRC. Spray on, then let soak in for a few minutes. Rub or scrape off the gasket. Use your fingernail or a plastic scraper to avoid damaging the cast-in raised bead which ensures sealing. Wash the part in solvent or water after using the gasket remover. Be careful where you spray because these are powerful paint removers!

Carb Spray Cleaner by STP is helpful for cleaning up carburetor and fuel-system parts. It's also useful for removing the residue and cleaning parts after using spray-on gasket remover.

A lug or protrusion on the metering block is a good prying point when the gasket is stuck. Avoid penetrating the gasket more than 1/4 in. because you can chisel off hidden locating tabs or dowels.

while looking down to see whether the throttle plates are actually vertical (wide-open position).

37. Throttle plates not seated in the bores. They're not supposed to seat in the bores! They're factory-adjusted to be slightly open at idle. If the throttle plates have shifted or have become loose, then you *may* want to consider recentering them. This is not a job we recommend for the amateur.

It requires removing the throttle body and reseating the plates. And it is a ticklish operation on most carburetors because throttle-plate screws are staked to prevent loosening. Care is needed to prevent twisting off the screw heads. Remove the upset or staked material with a small file and remove the screws with great care. Back off the idle-speed screw, reset the plates in the bores and retighten the screws. Support the oppo-

site side of the shaft and restake with a small prick punch.

IMPORTANT: These screws *must* be restaked to avoid screws and/or plates entering the engine to cause serious damage.

38. Dashpot binding or misadjusted. The dashpot slows the throttle return. If the idle-speed screw never returns to the stop, the dashpot is binding or misadjusted. Normally, the dashpot should have approximately 3/32-in. additional travel beyond closed throttle. Readjustment can be made by turning the dashpot after the locknut has been loosened. Be sure to tighten the locknut after the readjustment.

39. Vehicle linkage binding. Separate the vehicle throttle linkage from the carburetor to isolate the problem. Look for sticking or binding along the linkage. Lubrication may solve the problem. Cable-operated linkages sometimes require minor cable rerouting to avoid binding.

40. Anti-dieseling solenoid incorrectly adjusted. This solenoid acts as the idle stop while the engine is running. It is withdrawn when the igni-

tion is turned off so the plates can close farther in the bore to prevent after-run or dieseling. Make sure the solenoid plunger moves as the ignition is turned on and off. Idle speeds should be set with the solenoid *activated* (holding throttle slightly away from closed position).

41. Fast-idle cam bound or stuck. Squirting a little solvent on and around the cam and its associated linkage may correct this.

42. Overstressed throttle-linkage return spring. Check the spring from the carburetor throttle lever to some mounting point on the engine. If coil separation indicates an overstressed condition, replace the spring.

43. Bound or bent throttle shaft. Separate the throttle-actuating linkage from the carburetor throttle lever and check for binding or sticking. If deposits and dirt are the problem, squirting a little solvent on the shaft may help. A bent shaft will probably have to be replaced. Binding is sometimes caused by over-torquing carburetor-attachment nuts, warping the throttle body. This can prevent

Holley Renew Kits contain original parts. Many other kits do not and the parts may not perform to the desired specifications.

It's a good idea to place the carburetor on a stand while disassembling to protect the throttle plates. Many good ones are available, but four 5/16 bolts and eight nuts cost less than a dollar and work just great.

Open throttles extend below throttle body. Handle carburetor with care when it is off of the manifold. Avoid letting your friends play with it! Damage inevitably occurs when throttles are held open—then the carburetor is set down hard. You won't be able to run the carburetor until you've purchased new throttle plates—at your expense.

Don't even think about reusing this gasket! Holley's fuel bowl and metering block gaskets seal well, but they self-destruct on disassembly. Let the parts soak in lacquer thinner for a couple of hours. Or, remove the residue with spray-on gasket remover, time and patience. Be sure to blow out all passages where gasket pieces could create problems. Be extremely careful not to scrape off the cast-in raised-bead sealing surfaces!

diaphragm-operated secondary throttles from opening.

44. Worn throttle-shaft bearings. Separate the vehicle throttle linkage from the carburetor and wiggle throttle shaft up and down. If movement is noticeable, shaft bearings are worn. Replace the throttle body or have it rebushed by a carburetor rebuilder.

45. Failed secondary operating diaphragm. Remove the cover and inspect the diaphragm. Replace the diaphragm if it is torn or broken. Diaphragms are available as a service item. Part numbers are in the *Holley Performance Parts Catalog* or the *Holley Illustrated Parts & Specs Manual.*

46. Plugged secondary vacuum-signal port. This can be repaired by cleaning, but requires disassembly of the carburetor.

47. Binding secondary-throttle shaft. This shaft must be free to rotate to func-

tion properly. Over-torqued carburetor hold-down nuts or capscrews or using a thick soft gasket may result in throttle-body warping and hence throttle binding. Another, less-common cause may be loose or shifted throttle plates.

48. Secondary operating vacuum hose leaky or disconnected. This applies only to 3 x 2 (three-carburetor) applications.

49. Flooding. This is the condition when fuel continues to enter the carburetor uncontrolled by the float and inlet valve. Fuel spills out of the carburetor into the intake manifold. Look for leaky or worn fuel-inlet valves or foreign material holding the valve off the seat.

Foreign material is often rust or dirt from the vehicle or gasoline station fuel tank. Flooding can also be caused by a faulty float. Brass floats may develop a leak and lose buoyancy. This seldom happens with closed-cellular or hollow-plastic floats, although they too can absorb fuel over time and lose their buoyancy.

50. Broken power valve diaphragm. Remove the power valve and inspect. A backfire can be the cause. If you find this, be sure to install power-valve-blowout protection, see page 55.

51. Slow choke come-off. Can be caused by a choke misadjustment. A more common cause with hot-air chokes is clogging or rupturing of

151

the tube bringing hot air from the stove in the exhaust manifold or crossover to the choke unit. If hot air doesn't get to the choke bimetal, the choke will open late or remain in a partially closed position. On a completely warmed engine, the integral-choke unit should be quite hot to the touch.

52. Wrong metering-block gasket. Check to make sure you have the correct one. Note if it should be used with a carburetor not having an accelerator-pump transfer tube.

PREPARING FOR CARBURETOR REPAIR

Once you've analyzed the problem and decided it lies with the carburetor, you've set yourself up for *another* decision. Here are your options:
1. Take the carburetor apart, make the critical repair and put it back together.
2. Make an economy carburetor repair.
3. Perform a complete carburetor disassembly and repair.

Which Option?–If you've done a little carburetor work and you're fairly sure where the problem is and time is important, option 1 is probably the best approach. You might be able to perform this operation without removing the carburetor, but it isn't really a good idea. Bending over the fender is difficult and parts are easily lost.

If time allows, remove the carburetor from the engine. For this option you need, at the bare minimum, new gaskets. Gasket numbers for Model 2300, 4150/60/80 and 4165/75 are in the *Performance Parts Catalog* or in the *Holley Illustrated Parts & Specs Manual* (Holley Part 36-51-6). Your dealer probably has both. Also see pages 89—90. For other carburetors you may have to purchase a complete Renew Kit to obtain the gaskets you need.

The catalog and manual contain a wealth of information about specific Holley carburetors and can be obtained by writing:

Technical Service Department
Holley Replacement Parts Division
2800 Griffin Drive
Bowling Green, Kentucky 42101
(502) 781-9741

There is another *Holley Illustrated Parts & Specs Manual for Older Carburetor Models.* The Technical Service Department is also a good place to get quick service on hard-to-find parts that may not be immediately available at your dealer.

We think option 2—the economy carburetor repair method—represents the best compromise of cost, time requirement, and satisfactory results. Cleaners such as kerosene, Stoddard solvent and mild paint thinners allow you to clean up delicate plastic and synthetic parts together with metal parts, making complete disassembly unnecessary.

Cleaning–A small open can, a brush, a small scraping tool and some elbow grease let you do a fine job of washing. For dissolving deposits, use lacquer thinner, toluol, MEK, Gum-Out, Chem-Tool, STP Carb Cleaner or any of the many brands of cleaners available. These must be used in a well-ventilated area away from fire or pilot lights–such as on a gas range, water heater or furnace. Use with a brush on large surfaces to dissolve deposits. Use a common ear syringe to shoot the solution through passages. Wear safety glasses and keep your hands out of any dissolving-type chemicals. One way or another you can do a good job without exotic equipment.

Passages and orifices should be blown out with compressed air. If compressed air isn't available, you can always use a bicycle or tire pump to do the job. Never use a wire or drill to clean orifices and restrictions because the slightest mark can change flow characteristics.

Kits–As far as kits are concerned there are two ways to go, a Renew Kit or a Trick Kit. A Renew Kit is a repair kit with everything you'll need to repair your particular carburetor. It contains all gaskets including the one that goes between the carburetor and intake manifold. Also included are fuel-inlet valves, power valves, accelerator-pump diaphragms or cups and a service instruction sheet

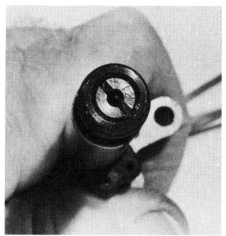

Feedback-type carburetors have a removable solenoid. Do not immerse this solenoid in strong carburetor cleaners! Don't remove the screw-in jet.

While the feedback carburetor is still on the car, put your hand on the feedback solenoid. You should be able to feel it working. If you can feel the vibrations, it is probably OK.

Eastwood makes this product to restore the factory-new finish to Holley cast parts. A good finish for your repair job.

giving adjustment specifications and procedures. If you run the same carburetor most of the time, get two and set one aside as a spare. This can be a real timesaver! Renew Kits are the most commonly used.

Trick Kits have the same items as Renew Kits, plus a selection of vent baffles, a pump-cam assortment, pump discharge nozzle assortment and where needed, an assortment of secondary springs. These kits are intended for those who are serious about tuning.

If you already have your carburetor dialed-in, you probably don't need a Trick Kit because it is considerably more expensive. If you are going to do quite a bit of tuning, assemble your own combination of parts and pieces.

When similar gaskets or parts are included, compare with the original parts and choose the parts most similar to the original. Parts will probably be left over from your kit when you've finished. Don't panic! The kits

were designed to service more than one application in most cases. "Consolidation" reduces the number of kits the dealer has to stock. It is actually more economical to include a few extra parts so each kit will service several different carburetors.

Tools—Common tools are usually good enough for carburetor disassembly and repair. You will need standard and Phillips screwdrivers, regular and needle-nose pliers and the usual set of open-end wrenches. A sharp scraping tool is good for removing deposits as well as old gaskets from the intake manifold.

Be extremely careful with gasket surfaces on carburetor parts. Zinc and aluminum are easily scratched. Small "beads" or raised surfaces may be cast in to provide positive sealing. Because they are hidden under stuck-on gasket material, they're easily damaged.

You can either soak the parts in lacquer thinner (not plastic parts!) or

use spray-on gasket remover. Take your time and use several applications if that's what it takes to loosen the old gasket material so you can get it off with a soft plastic scraper–or your fingernail.

A special 1-in. open-end wrench (MAC S-141) may be required to remove the fuel-inlet fitting on Model 2300, 4150/60/80 and 4165/75 carburetors. You can sometimes do this with a standard open-end wrench. Corbin-clamp pliers and tubing wrenches are good to have for carburetor removal and installation. Tubing wrenches are a *necessity* if you are working with soft fuel-line nuts.

Follow the instructions in Chapter 4, *Select & Install Your Carburetor* for removing and installing the carburetor. There's no way to overstress the importance of tagging and identifying each vacuum hose with a piece of tape or other tag. Take a few minutes to make a schematic diagram showing all of the hookups. This saves a lot of grief, especially on vehicles with complicated emission-control systems.

A holding fixture should be used to prevent damage to the throttle plates while working on the carburetor. Many stands are available, but 5/16 bolts and nuts work just fine. Use a second nut on each bolt to lock the throttle body onto the bolts, preventing wobble.

Now you're ready for the job. Specific carburetor sections include disassembly, repair and assembly procedure for typical performance carburetors. In some cases, we discussed the complete disassembly even though you may be using the economy method.

We couldn't include every variation of every model carburetor Holley ever made. Expect to do a little thinking and analysis of your own application from the basics presented here.

Models 4150/4160/4180, 4165/4175, 2300 Repair & Adjustment

> **NOTE:** The Model 4150 family of carburetors includes Models 4160, 4180, 4165, 4175, and 2300. Because of similar construction they are discussed in one set of disassembly, repair and adjustment procedures. The Model 4500 is similar, so these instructions also apply to those carburetors.

For most photos we used a popular Model 4150 which incorporates more sub-assemblies than most carburetors. A few photos show construction peculiar to specific carburetors. If you can make it through this one, you won't have any problem with any other carburetor in this family.

(1) Before beginning disassembly, mount the carburetor on a holding fixture or a set of 5/16-in. bolts. Loosen the fuel-inlet fitting/s now and also the fuel-bowl sight plugs and needle and seat lock screws because it is easier to do while the carburetor is completely assembled. I prefer to loosen these parts while the carburetor is still securely bolted to the manifold.

(2) Remove 4 secondary fuel-bowl screws. Remove fuel bowl and gasket and the metering block and its gasket. Separating assemblies may require a rap from screwdriver handle or plastic mallet. Separation may be difficult on carburetors with black gaskets. If inserting a putty knife or anything else to loosen gasket, don't push in more than 1/4 in. or you may shear off a locating dowel.

This shows a Model 4150 with dual inlets and no fuel-transfer tube between bowls.

(3) Single-inlet Model 4160 (and 4165 and 4180) has balance tube between bowls. Don't worry about preserving O-ring (arrow) if you are rebuilding—new ones are in Renew Kit. Just slide fuel bowls off of tube. On reassembly, use petroleum jelly or oil on new O-rings. Twist tube slightly to ease O-ring entry into each bowl assembly. Make sure no part of O-ring gets pinned over bead of tube. Fuel under pressure is contained in tube, so don't make mistakes or you could cause leaks.

(4) Now repeat the procedure on primary side. On Model 2300 two-barrel this is where you begin. Use plenty of torque when replacing the bowl screws: 50 in. lb. Bowl screws were originally slotted. In 1987 they were changed to 5/16-in. hex-head flanged capscrews.

(4A) Removing secondary bowl on 4160 with dual fuel inlets reveals secondary metering plate attached with clutch-head screws.

Model 4160, 4165 and 4180 carburetors have a metering plate instead of a metering block. These also tend to stick after removing the clutch-head screws. A rap should loosen it.

(5) Next remove the choke unit. With integral chokes, the first thing to do is to remove the little hairpin retainer from the bottom of the rod connecting the choke-control—through the main body casting—up to the choke lever. Use needle-nose pliers.

(6) Before going further, note the position of the mark on the black bimetal housing relative to marks on the choke casting. The strut in the middle of these marks on the casting is called the *index* mark. Choke adjustments are designated relative to this mark. Draw a picture so you can reassemble the choke in the same position. Choke settings are called out in marks RICH or LEAN from the index. Consult the *Holley Illustrated Parts & Specs Manual* if you've lost or forgotten the setting.

(7) Remove 3 screws holding the bimetal housing retainer and remove retainer, gasket and bimetal housing. Do not remove the bimetal from the housing.

(8) The metal choke housing can be removed from the main body by the removal of three attachment screws.

(9) For carburetors with a divorced or remote choke, remove 2 screws holding the vacuum break, the retainer holding the fast-idle cam and lever and the choke rod retainer if there is one. Disconnect vacuum hose at the throttle body and remove all of the choke parts. This photo shows the choke lever and vacuum break from a typical divorced or remote-choke application.

(10) Remove clip from shaft (arrow) and take out 3 screws mounting vacuum secondary housing to main body. Remove vacuum secondary housing.

155

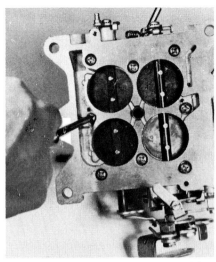

(11) Take out 6 or 8 screws attaching throttle body to main body. These are often tough to turn. Impact driver may be needed. Two center screws are not used on later carburetors, though main body is tapped to allow their use.

(12) Center-pivot-float fuel bowls hinge the primary floats at front and secondary floats at back. Bowl can have either single or dual inlets. With dual inlets, one side is closed with a plug and the other side has an inlet fitting. Plastic float is shown here. Earlier carburetors had brass floats shown in (13) at right.

(13) On bowls with front-mounted floats, first remove the inlet needle-and-seat assembly. Loosen the lock screw and turn the hex nut counter-clockwise. The hex nut slips over the needle-and-seat assembly. Now remove 2 screws holding the float-mounting bracket and remove the bracket and float from the fuel bowl.

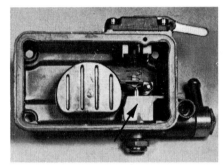

(14) Fuel bowl with side-mounted float. Remove baffle (arrow) surrounding inlet valve and then retainer from float hinge pin. Float assembly can then be removed. Non-adjustable inlet valves must be removed from the inside with an open-end wrench. Take out fuel-inlet fitting you previously loosened. Remove it with the integral filter and spring (see photo). Remove sight plugs if used. The fuel bowls are completely disassembled. This fuel bowl has an externally adjustable needle/seat and sight plug.

(15) Disassembled side-mounted-float fuel bowl with a non-adjustable inlet valve.

(15A) Plastic float in side-mount fuel bowl with externally adjustable inlet valve. Note bumper spring to stabilize float.

(16/17) Primary bowls (and secondary bowls on double-pumpers) contain accelerator-pump assemblies. Remove 4 attachment screws and lift pump diaphragm and housing from the bowl. Typical rubber-umbrella inlet valve is in top photo. Remove rubber valve before putting fuel bowl in cleaner

Older carburetors used a hanging-ball inlet (lower photo). Clearance between ball and retainer (arrow) should be 0.011-0.013 in. with bowl inverted.

(18) Remove main metering jets with a wide-blade screwdriver. The screwdriver must cover both sides of the slot or you will damage the jets. Primary and secondary jets are often different sizes. Small jets go in the primary side if the carburetor has smaller primary venturis. On rare occasions jets will differ from side to side in the same metering block. Always check jet sizes and locations and write this down before you take out the jets. Remove bowl-vent splash shield and any vacuum fittings if used. **DO NOT REMOVE ANY PRESSED-IN VACUUM TUBES.** Count the turns (and part turns) as you turn each mixture screw in gently until it seats. Record the settings. Take out idle-mixture screws and seals.

EXCEPTION: Model 4180 mixture screws are sealed in the throttle body. It is almost impossible to unseal and remove them, so this is not recommended for cleaning/rebuilding.

Remove power valves with a 1-in. wrench. A socket wrench is preferred, but an open end is OK if you proceed carefully. The 1-in. wrench for large

fuel-inlet fittings can be used.

It is not usually necessary to disassemble metering blocks any further. Although there are tubes inside the main and idle wells, these can usually be cleaned with compressed air and carburetor solvent/cleaner. The small metering plate from the secondary side of Model 4160 requires no further disassembly.

Metering blocks with an O-ringed tube connecting them to the main body should have this tube removed before proceeding further.

Use new gaskets on reassembly. When you reassemble the carburetor, install the mixture screws 1-1/2 turns off their seats as a starting point.

(19/20) Take a good look at this assembly before taking it apart. Note the fast-idle cam and choke lever relationship. Remove the choke shaft nut, lock washer and spacer and then slide the shaft and fast-idle cam from the housing. Next remove the choke qualifying piston. Make sure it operates freely in its bore. Remove the cork gasket that surrounds the restriction on the back side of the choke housing. Use a new gasket on reassembly.

(20A) Details of electric choke caps. Older model at top is heated by a resistance wire (arrow). Lower one uses solid-state heater that raises resistance and lowers current flow as temperature rises.

CLEANING & ASSEMBLY

Now your carburetor is completely disassembled and ready for cleaning. Remember if you've chosen the complete carburetor repair method, *only metal parts should be immersed in the special cleaner.* Do not immerse electrical parts in strong cleaners. All non-metal parts including the choke bimetal and housing, Teflon bushings and the plastic accelerator-pump cams should be cleaned with a milder cleaner such as kerosene or paint thinner. Once all cleaning is done, inspect the parts for excessive wear to determine if they should be replaced. If you've purchased the Renew Kit, all gaskets, pump diaphragms, secondary diaphragm (in some cases), fuel-inlet valves and power valves are included.

(21) Remove 4 attachment screws attaching cover to vacuum-secondary housing. A rap will separate the upper and lower housings. The spring and diaphragm can now be removed. My pencil points to a cork gasket that should be removed. Use a new gasket on reassembly.

(22) If you plan to immerse the main body in a carburetor cleaner, remove the choke plate and shaft to remove the little plastic guide. If you use the economy repair method, choke-shaft removal is seldom necessary and we don't recommend it. If you plan to remove the choke plate and shaft, first file the staked portion of the choke screws. Then take out the screws and lift the choke plate out of the shaft. The choke shaft can now be removed from the main body, allowing the removal of the choke rod and guide. Remember, the choke-plate screws must be restaked upon reassembly. This requires supporting the shaft so it will not be bent.

Remove the pump-discharge-nozzle screw, the nozzle and gasket. Turn the body over to remove the pump-discharge valve (weight or ball). Double pumpers have two pump-discharge assemblies. Models 4165/4175 do not have a pump-discharge valve at this location.

(23) It is seldom necessary to disassemble the throttle body and immerse all of the parts. First of all, only a few metering restrictions and small passages are in the throttle body. Second, throttle-plate screws must be removed and restaked upon assembly. This adds up to a real job and if it isn't necessary, why do it? Cleaning with a brush and a mild solvent is usually more than adequate. If you insist on a complete disassembly, your work will include: Remove the idle-speed screw and spring, remove the diaphragm-operating lever from the secondary throttle shaft and the fast-idle lever from the primary shaft, remove the cotter key and the connecting link between the primary and secondary throttle levers. File off staked ends of the throttle-plate attachment screws, remove screws and throttle plates. Slide the shafts out of the flange. Take out the Teflon bushings (typical on secondary side of carburetors with diaphragm-operated secondaries).

REASSEMBLY

In reassembling the carburetor, simply follow the disassembly instructions in reverse order. Pictures and exploded views in this book will be very helpful.

There are a few things to look for:

When assembling the fuel-supply tube, use a little petroleum jelly (Vaseline) on the O-rings so they'll slip in easily. When assembled properly, you should be able to rotate the tube with your fingers.

Be sure to do a good job of restaking the choke and throttle-attachment screws. If one of these screws comes loose and goes into the engine, it could be very expensive.

Other than the very small choke and throttle screws, don't be afraid to torque the assembly screws, especially the throttle-body and fuel-bowl screws. Once you've gotten them all in, it's a good idea to go around again and give them a little extra.

On Models 4165/4175 the short bowl screws go on top.

Don't forget to replace all of the gaskets. Some little ones are easy to overlook.

(24) Be sure the secondary diaphragm is sealed all the way around when you attach the cover to the lower housing. Unload the diaphragm spring by pushing up on the rod. This will cause the diaphragm to lay flatter and make it easier to get the screws and cover installed.

(25) Some bimetals have a hooked end; others have loop as shown. Capture the choke-lever tang with the bimetal end when installing the housing. Rotate bimetal housing back and forth before tightening retainer—choke plate should move as you do this.

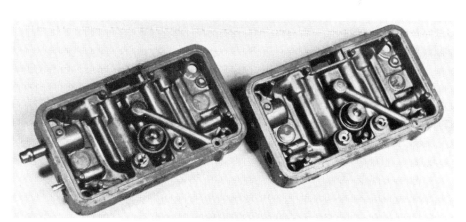

(26) Most primary metering blocks can be distinguished from the secondary because they contain the idle-mixture screws and may also contain one or more vacuum tubes. However, some race carburetors have idle screws in the secondary metering blocks. If you have one block without a power valve, that will be the secondary metering block. Metering-block gaskets are properly installed when they line up with the little dowels on the metering block. Be sure to use the correct metering block to carburetor body gaskets, see pages 89-90.

ADJUSTMENTS

A number of adjustments should be made as you reassemble the carburetor. Adjustment procedures are shown in the accompanying photographs. Specific dimensions are in the instruction sheet supplied with Renew Kit and in the *Holley Illustrated Parts & Specs. Manual.*

(27) Dry float setting is usually measured between the bowl casting and the float end with bowl inverted. Holley Renew Kits give correct specs and may include a gauge. The *Holley Illustrated Parts & Specs Manual* is another good source. Adjust float level by bending tang as shown, being careful not to mar the contact surface.

(28) Choke-qualify adjustment. Specification is clearance between choke plate and casting on either top or bottom edge as noted. Repair kit or *Holley Illustrated Parts & Specs Manual* gives dimension. Hold choke plate closed and measure clearance while operating vacuum break by hand or with a vacuum source. A vacuum source is best—your engine is an excellent one. This can be done after the carburetor is installed. Adjust by bending vacuum break link on divorced-choke carburetors.

(29) Dechoke spec is measured between choke plate and housing with throttle wide-open. Adjust by bending end of choke rod as indicated by screwdriver.

(30) Emission-type vent valve connects to charcoal canister. Clearance should be 0.015 in. as shown with throttle at normal or curb idle. Adjust by bending lever.

(31) Pump lever should always have at least 0.015- to 0.020-in. additional travel beyond screw when throttle is wide-open. Screw and lever should also be in contact at idle. Adjust screw to accomplish both. With a green cam, added pump-lever travel should be 0. Pump cam is shown in position 2 (arrow). To reduce capacity and change delivery, remove screw, move cam and insert in hole 1 and screw into alternate hole in cam.

(32) Old-style external vent clearance is usually around 3/32 in. Adjust by bending lever.

(33) Normal or curb idle is set as shown. Set it to correct engine rpm later.

(34) Secondary idle-speed adjustment. About 1/2 to 1 turn away from having plates seated in the bore is a good initial setting.

(35) Fast-idle adjustment is made with fast-idle screw on highest step of fast-idle cam. Clearance between primary throttle plate and the throttle bore should be about 0.025 in., measured as shown.

(36) Dashpot setting refers to additional travel of dashpot when throttle is at curb idle. 0.090—0.120 in. is common. Adjust by loosening locknut and turning dashpot assembly. Consult *Holley Illustrated Parts & Specs Manual* or Renew Kit instruction sheet for exact setting.

(37) Externally adjustable fuel-bowl float levels can be set *on the vehicle*. Remove the sight plug. Loosen lock screw at the top of the assembly and turn adjusting nut until fuel level is at bottom of the sight-plug hole. To confirm setting, flush fuel bowl a few times by accelerating the engine with transmission in neutral. Tighten lock screw while holding adjustment nut and replace sight plug. This operation is difficult to do accurately on a rough-idling vehicle.

MODEL 2300 WARNING!
Model 2300 throttle bores ARE NOT CENTERED. They are offset toward rear of carburetor base. Careless gasket installation can prevent correct throttle operation. The engine may start and run, but throttles may snag on gasket and not close.

MODEL 2300

for 3 x 2-barrel setup
Holley Typical View 37—1

Center carburetor
with accelerator pump
and choke

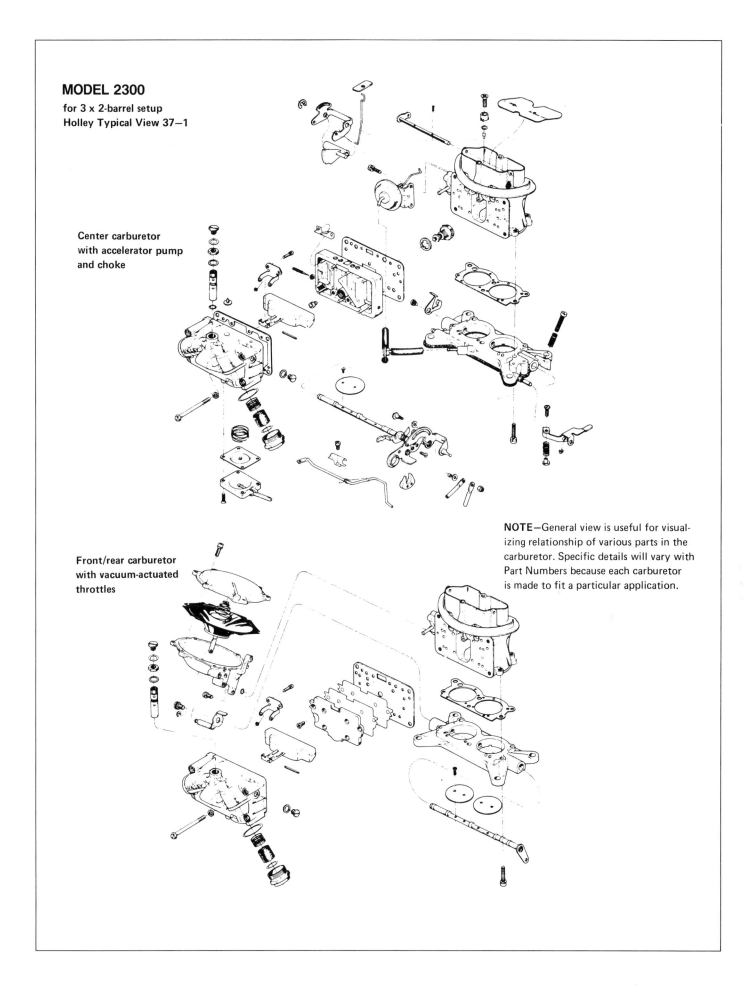

Front/rear carburetor
with vacuum-actuated
throttles

NOTE—General view is useful for visualizing relationship of various parts in the
carburetor. Specific details will vary with
Part Numbers because each carburetor
is made to fit a particular application.

MODEL 2300
Holley Typical View 3-3

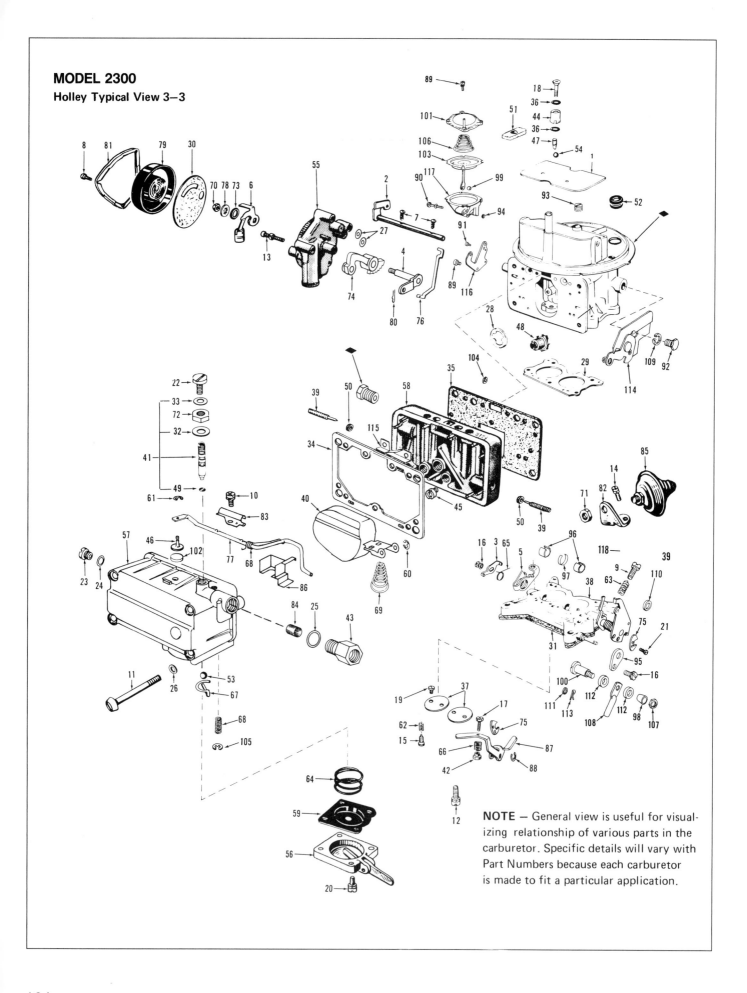

NOTE — General view is useful for visualizing relationship of various parts in the carburetor. Specific details will vary with Part Numbers because each carburetor is made to fit a particular application.

1	Choke plate
2	Choke shaft assembly
3	Fast idle pick-up lever
4	Choke hsg. shaft & lev. assy.
5	Fast idle cam lever
6	Choke therm. lev., link & piston
7	Choke plate screw
8	Therm. hsg. clamp screw
9	Throttle stop screw
10	Air vent rod clamp scr. & LW
11	Fuel bowl to main body screw
12	Throt. body scr. & LW
13	Choke hsg. scr. & LW
14	Dashpot brkt. scr. & LW
15	Fast idle cam lever screw
16	Fast idle cam lev. & throt. lev. screw & LW
17	Pump oper. lev. adj. screw
18	Pump discharge nozzle screw
19	Throttle plate screw
20	Fuel pump cov. assy. scr. & LW
21	Pump cam lock scr. & LW
22	Fuel valve seat lock screw
23	Fuel level check plug
24	Fuel level check plug gasket
25	Fuel inlet fitting gasket
26	Fuel bowl screw gasket
27	Choke housing gasket
28	Power valve body gasket
29	Throttle body gasket
30	Choke therm. housing gasket
31	Flange gasket
32	Fuel valve seat adj. nut gskt.
33	Fuel valve seat lock scr. gaskt.
34	Fuel bowl gasket
35	Metering body gasket
36	Pump discharge nozzle gasket
37	Throttle plate
38	Throt. body & shaft assy.
39	Idle adjusting needle
40	Float & hinge assy.
41	Fuel inlet valve & seat assy.
42	Pump oper. lev. adj. scr. fitting
43	Fuel inlet fitting
44	Pump discharge nozzle
45	Main jet
46	Air vent valve
47	Pump discharge needle valve or check ball weight
48	Power valve assembly
49	Fuel valve seat "O" ring seal or gasket
50	Idle needle seal
51	Choke rod seal
52	Choke cold air tube grommet
53	Pump inlet check ball
54	Pump discharge check ball
55	Choke hsg. & plugs assy.
56	Fuel pump cover assy.
57	Fuel bowl & plugs assy.
58	Main metering body & plugs assy.
59	Pump diaphragm assembly
60	Float spring retainer
61	Air vent retainer
62	Fast idle cam lev. scr. spring
63	Throttle stop screw spring
64	Pump diaphragm return spring
65	Fast idle cam lev. spring
66	Pump oper. lev. adj. spring
67	Pump inlet check ball retainer
68	Air vent rod spring
69	Float spring
70	Choke thermostat shaft nut
71	Dashpot screw nut
72	Fuel valve seat adj. nut
73	Choke therm. lever spacer
74	Fast idle cam assembly
75	Pump cam
76	Choke rod
77	Air vent rod
78	Choke therm. shaft nut LW
79	Thermostat housing assembly
80	Choke rod retainer
81	Thermostat housing clamp
82	Dashpot bracket
83	Air vent rod clamp
84	Filter screen assembly
85	Dashpot assembly
86	Baffle plate
87	Pump operating lever
88	Pump operating lev. retainer
89	Adapter mounting & diaphragm cover assy. screw
90	Throt. diaphragm hsg. scr.
91	Adapter passage screw
92	Choke bracket screw
93	Air adapter hole plug
94	Throt. diaphragm hsg. gasket
95	Throttle lever
96	Throttle shaft bearing
97	Throttle shaft brg. (center)
98	Throttle connector pin bushing
99	Diaphragm check ball
100	Throttle connector pin
101	Diaphragm housing cover
102	Air vent cap
103	Diaphragm housing assembly
104	Diaphragm link retainer
105	Air vent rod spg. retainer
106	Diaphragm spring
107	Throttle link connector pin nut
108	Throttle connector bar
109	Choke brkt. scr. lock washer
110	Throt. link connector pin washer
111	Throt. connector pin washer
112	Throttle connector pin spacer
113	Throt. connector pin retainer
114	Choke control lever bracket
115	Metering body vent baffle
116	Throt. diaphragm adapter
117	Diaphragm housing
118	Idle adj. needle spring

MODEL 4150/4160
(also 3150/3160)

Holley Typical View 24–1

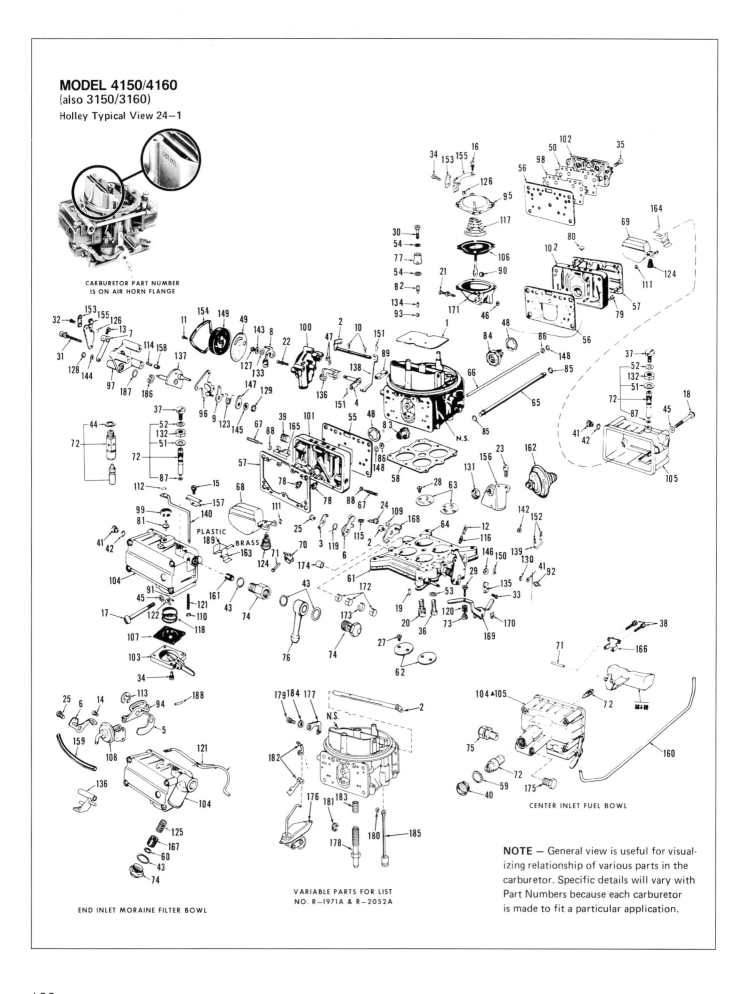

CARBURETOR PART NUMBER
IS ON AIR HORN FLANGE

END INLET MORAINE FILTER BOWL

VARIABLE PARTS FOR LIST
NO. R–1971A & R–2052A

CENTER INLET FUEL BOWL

NOTE — General view is useful for visualizing relationship of various parts in the
carburetor. Specific details will vary with
Part Numbers because each carburetor
is made to fit a particular application.

1 Choke plate
2 Choke shaft assembly
3 Fast idle pick-up lever
4 Choke housing shaft & lever assy.
5 Choke control lever
6 Fast idle cam lever
7 Choke lever & swivel assy.
8 Choke therm. lev., link & piston assembly
9 Choke rod lev. & bush. assy.
10 Choke plate screw
11 Therm. housing clamp screw
12 Throttle stop screw
13 Choke lever assembly swivel screw
14 Choke diaph. assy., brkt. scr. & lock washer
15 Air vent clamp screw & LW
16 Sec. diaph. assy. cov. scr. & LW
17 Fuel bowl to main body screw — primary
18 Fuel bowl to main body screw — secondary
19 Diaph. lever adjusting screw
20 Throt. body screw & lock washer
21 Diaph. hsg. assy. scr. & LW
22 Choke housing screw & LW
23 Dashpot brkt. screw & LW
24 Fast idle cam lever adj. screw
25 Fast idle cam lev. scr. & LW
26 Diaph. lev. assy. scr. & LW
27 Throt. plate screw — primary
28 Throt. plate screw — secondary
29 Pump lever adjusting screw
30 Pump discharge nozzle screw
31 Fast idle cam plate scr. & LW
32 Choke cont. wire brkt. clamp scr.
33 Pump cam lock screw
34 Fuel pump cov. assy. scr. & LW
35 Secondary metering body screw
36 Throt. body screw — special
37 Fuel valve seat lock screw
38 Float shaft brkt. scr. & LW
39 Spark hole plug
40 Fuel bowl plug
41 Fuel level check plug
42 Fuel level check plug gasket
43 Fuel inlet fitting gasket
44 Fuel valve seat gasket
45 Fuel bowl screw gasket
46 Sec. diaphragm housing gasket
47 Choke housing gasket
48 Power valve body gasket
49 Choke thermostat housing gasket
50 Sec. metering body plate gasket
51 Fuel valve seat adj. nut gasket
52 Fuel valve seat lock screw gasket
53 Throt. body screw gasket
54 Pump discharge nozzle gasket
55 Metering body gasket — primary
56 Metering body gasket — secondary
57 Fuel bowl gasket
58 Throttle body gasket
59 Fuel bowl plug gasket
60 Fuel inlet filter gasket

61 Flange gasket
62 Throttle plate — primary
63 Throttle plate — secondary
64 Throt. body & shaft assembly
65 Fuel line tube
66 Balance tube
67 Idle adjusting needle
68 Float & hinge assy. — primary
69 Float & hinge assy. — secondary
71 Float lever shaft
72 Fuel inlet valve & seat assy.
73 Pump lever adjusting screw fitting
74 Fuel inlet fitting
75 Fuel transfer tube fitting assy.
76 Fuel inlet tube & fitting assy.
77 Pump discharge nozzle
78 Main jet — primary
81 Air vent valve
82 Pump discharge needle valve
83 Power valve assy. — primary
85 Fuel line tube "O" ring seal
86 Balance tube "O" ring seal
87 Fuel valve seat "O" ring seal
88 Idle needle seal
89 Choke rod seal
90 Diaphragm housing check ball — sec.
91 Pump inlet check ball
92 Throttle lever ball
93 Pump discharge check ball
94 Choke diaphragm assembly link
95 Sec. diaphragm housing cover
96 Back-up plate & stud assembly
97 Fast idle cam plate
98 Secondary metering body plate
99 Air vent cap
100 Choke hsg. & plugs assembly
101 Main metering body & plugs assy. — primary
102 Main metering body & plugs assy — secondary
103 Fuel pump cover assembly
104 Fuel bowl & plugs assy. — primary
105 Fuel bowl & plugs assy. — secondary
106 Secondary diaph. & rod assy.
107 Pump diaphragm assembly
108 Choke diaphragm assembly — complete
109 Secondary diaph. link retainer
110 Air vent rod spring retainer
111 Float retainer
112 Air vent valve retainer
113 Choke control lever retainer
114 Fast idle cam plunger spring
115 Fast idle cam lever screw spring
116 Throttle stop screw spring
117 Secondary diaphragm spring
118 Diaphragm return spring
119 Fast idle cam lever spring
120 Pump lev. adj. screw spring
121 Air vent rod spring
122 Pump inlet check ball ret. spring
123 Choke spring
124 Float spring — pri. & sec.
125 Fuel inlet filter spring
126 Choke cont. wire brkt. clamp scr. nut
127 Choke thermostat shaft nut

128 Choke control lever nut
129 Back-up plate stud nut
130 Throttle lever ball nut
131 Dashpot nut
132 Fuel valve seat adj. nut
133 Choke thermostat lever spacer
134 Pump check ball weight
135 Pump cam
136 Fast idle cam assembly
137 Fast idle cam & shaft assembly
138 Choke rod
139 Throttle connecting rod
140 Air vent push rod
141 Throttle lev. ball nut washer
143 Choke shaft nut lock washer
144 Choke control lev. nut lock washer
145 Back-up plate stud nut lock washer
146 Throt. connector pin washer
147 Choke spring washer
148 Balance tube washer
149 Therm. hsg. assy. — complete
150 Throt. connector pin retainer
151 Choke rod retainer
152 Throt. connecting rod cotter pin
153 Choke cont. wire brkt. clamp
154 Thermostat housing clamp
155 Choke control wire bracket
156 Dashpot bracket
157 Air vent rod clamp
158 Fast idle cam plunger
159 Choke vacuum tube
160 Fuel transfer tube
161 Filter screen
162 Dashpot assembly
163 Baffle plate — primary (brass)
164 Baffle plate — secondary
165 Metering body vent baffle
166 Float shaft retainer bracket
167 Fuel inlet filter
168 Diaphragm lever assembly
169 Pump operating lever
170 Pump operating lever retainer
171 Secondary diaphragm housing
172 Throt. shaft bearing — sec (ribbon)
173 Throt. shaft bearing — sec. (ribbon)
174 Throt. shaft bearing — pri. (solid)
175 Fuel bowl drain plug
176 Choke assy. — complete (divorced)
177 Choke shaft lever
178 Idle by-pass adj. screw
179 Choke lever screw & LW
180 Choke piston link retainer
181 Fast idle cam retainer
182 Choke rod clevis clip
183 Idle by-pass adj. screw spring
184 Choke piston lever spacer
185 Choke piston & link assembly
186 Choke oper. lev. spring washer
187 Choke oper. lever washer
188 Choke clevis pin
189 Baffle plate — primary (plastic)

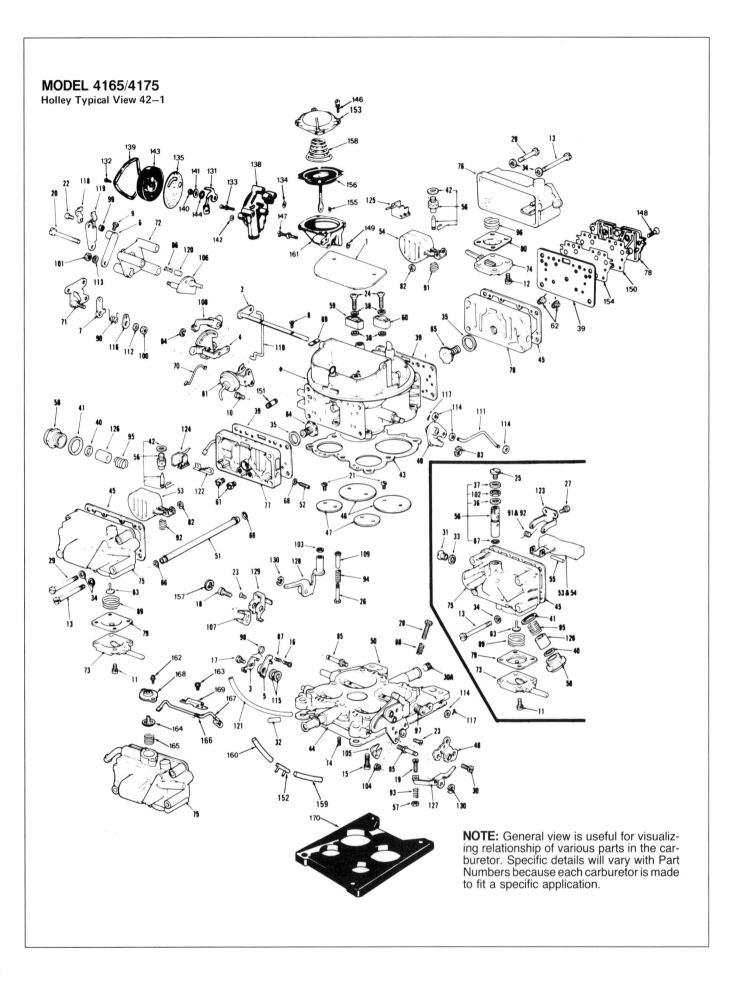

MODEL 4165/4175
Holley Typical View 42–1

NOTE: General view is useful for visualizing relationship of various parts in the carburetor. Specific details will vary with Part Numbers because each carburetor is made to fit a specific application.

1	Choke plate
2	Choke shaft assembly
3	Fast idle pick-up lever
4	Choke control lever
5	Fast idle cam lever
6	Choke lever & swivel assembly
7	Choke rod lever & bushing assembly
8	Choke plate screw
9	Choke lever swivel screw
10	Choke diaphragm bracket screw & LW
11	Fuel pump cover screw & LW — primary
12	Fuel pump cover screw & LW — secondary
13	Fuel bowl screw (long) pri. & sec.
14	Pump lever adjusting screw — secondary
15	Throttle body screw & LW
16	Fast idle cam lever adj. screw
17	Fast idle cam lever screw & LW
18	Pump cam lever screw & LW
19	Pump lever adj. screw — primary
20	Fast idle cam plate screw & LW
21	Throttle plate screw — pri. & sec.
22	Choke wire bracket clamp screw
23	Pump cam screw
24	Pump discharge nozzle screw
25	Fuel valve seat lock screw
26	Pump operating lever adj. screw
27	Float shaft bracket screw & LW
28	Throttle stop screw
29	Fuel bowl screw (short) pri. & sec.
30	Throttle lever extension screw
30A	Throttle body channel plug
31	Fuel level check plug
32	Vacuum tube plug
33	Fuel level check plug gasket
34	Fuel bowl screw gasket
35	Power valve gasket
36	Fuel valve seat adj. nut gasket
37	Fuel valve seat lock screw gasket
38	Pump discharge nozzle gasket
39	Metering body gasket — pri. & sec.
40	Fuel inlet filter gasket
41	Fuel inlet fitting gasket
42	Fuel valve gasket — pri. & sec.
43	Throttle body gasket
44	Flange gasket
45	Fuel bowl gasket - pri & sec.
46	Throttle plate — secondary
47	Throttle plate — primary
48	Throttle lever extension
49	Cam follower lever assy.
50	Throttle body & shaft assy.
51	Fuel line tubing
52	Idle adjusting needle
53	Float & hinge assy. — primary
54	Float & hinge assy. — secondary
55	Float shaft
56	Fuel inlet valve & seat assy.
57	Pump lever adj. screw fitting
58	Fuel inlet fitting
59	Pump discharge nozzle - primary
60	Pump discharge nozzle - secondary
61	Main jet — primary
62	Main jet — secondary
63	Pump check valve
64	Power valve assy - primary
65	Power valve assy — secondary
66	Fuel line tube "O" ring seal
67	Fuel valve seat "O" ring seal
68	Idle needle seal
69	Choke rod seal
70	Choke diaphragm link
71	Back-up plate & stud assy.
72	Fast idle cam plate
73	Fuel pump cover assy. - primary
74	Fuel pump cover assy — secondary
75	Fuel bowl & plugs assy — primary
76	Fuel bowl & plugs assy — secondary
77	Metering body & plugs assy — primary
78	Metering body & plugs assy — secondary
79	Pump diaphragm assy. — primary
80	Pump diaphragm assy. — secondary
81	Choke diaphragm assy.
82	Float hinge retainer
83	Cam follower lever assy. retainer
84	Choke control lever retainer
85	Pump lever stud
86	Fast idle cam plunger spring
87	Fast idle cam lever screw spring
88	Throttle stop screw spring
89	Diaphragm return spring — primary
90	Choke spring
91	Float spring — secondary
92	Float spring — primary
93	Pump lever adj. screw spring — primary
94	Pump lever adj. screw spring — secondary
95	Fuel inlet filter spring
96	Diaphragm return spring — secondary
97	Throttle return spring — secondary
98	Fast idle cam lever spring
99	Choke wire bracket clamp screw nut
100	Back-up plate stud nut
101	Choke lever nut
102	Fuel valve seat adj. nut
103	Pump operating lever adj. nut
104	Throttle lever ext. screw nut
105	Pump cam — primary
106	Fast idle cam & shaft assy.
107	Pump cam — secondary
108	Fast idle cam assy.
109	Pump operating lever screw sleeve
110	Choke rod
111	Secondary connecting rod
112	Back-up plate stud nut LW
113	Choke control lever nut LW
114	Secondary connecting rod washer
115	Throttle seal washer
116	Choke spring washer
117	Secondary connecting rod cotter pin
118	Choke wire bracket clamp
119	Choke wire bracket
120	Fast idle cam plunger
121	Choke vacuum hose
122	Metering body vent baffle — pri. & sec.
123	Float shaft retaining bracket
124	Baffle plate — primary
125	Baffle plate — secondary
126	Fuel inlet filter
127	Pump operating lever — primary
128	Pump operating lever & guide assy. — sec.
129	Pump cam lever — secondary
130	Pump operating lever retainer — pri. & sec
131	Choke therm. lever
132	Choke therm. cover screw
133	Choke housing screw
134	Choke housing gasket
135	Choke therm. cover gasket
136	Tube & "O" ring assy.
137	Idle adj. needle limiter cap
138	Choke housing & plugs assy.
139	Choke therm. cover retainer
140	Choke therm. shaft nut
141	Choke shaft nut lock washer
142	Choke housing screw & L.W.
143	Choke therm. cover assy.
144	Choke shaft spacer
145	Cam follower stud
146	Secondary diaphragm cover screws
147	Secondary diaphragm housing screws
148	Secondary metering body screws
149	Secondary diaphragm housing gasket
150	Secondary metering body plate gasket
151	Tube & "O" ring assy.
152	Four-way connector
153	Diaphragm cover machine
154	Secondary metering body plate
155	Secondary check valve
156	Secondary diaphragm
157	Secondary diaphragm link retainer
158	Secondary diaphragm spring
159	Choke vacuum hose
160	Choke vacuum hose
161	Secondary housing & seat assy.
162	Vent valve screws
163	Air vent rod clamp screw & L.W.
164	Vent valve body assy.
165	Vent valve spring
166	Vent rod spring
167	Vent valve rod
168	Vent valve clamp assy.
169	Air vent rod clamp
170	Throttle body spacer

MODEL 4180C

NOTE: General view is useful for visualizing relationship of various parts in the carburetor. Specific details will vary with Part Numbers because each carburetor is made to fit a specific application.

1 Choke plate
2 Choke shaft & lever assy.
3 Fast idle cam lever
4 Choke housing shaft & lever assy.
5 Choke thermostat lever & piston assy.
6 Choke plate screw
7 Choke thermostat clamp screw
8 Throttle stop screw
9 Pump cover screw and 1.w.
10 Secondary diaphragm cover screw & 1.w.
11 Fuel bowl screw (primary)
12 Secondary idle adjust screw
13 Throttle body screw & 1.w.
14 Secondary diaphragm housing screw & 1.w.
15 Fast idle cam lever & secondary diaphragm
   lever screw
16 Throttle plate screw
17 Fuel bowl screw (secondary)
18 Pump cam lock screw
19 Secondary metering body screw
20 Pump lever adjusting screw
21 Needle & seat lock screw
22 Modulator bracket screw
23 Choke housing screw & 1.w.
24 Fast idle adjusting screw
25 Transmission kickdown adjusting screw
26 Pump discharge nozzle screw
27 Fuel level check plug
28 Fuel bowl gasket
29 Power valve gasket
30 Needle & seat adjust nut gasket
31 Needle & seat lock screw gasket
32 Fuel inlet fitting gasket
33 Fuel inlet filter gasket
34 Secondary metering plate gasket
35 Secondary diaphragm & choke housing gasket
36 Fuel level check plug gasket
37 Fuel bowl screw gasket
38 Pump discharge nozzle gasket
39 Throttle body gasket
40 Secondary metering body gasket
41 Primary metering body gasket
42 Flange gasket
43 Choke thermostat gasket
44 Primary throttle plate
45 Secondary throttle plate
46 Primary throttle shaft assy.
47 Secondary throttle shaft assy.
48 Throttle body & shaft assy.
49 Throttle shaft bushing
50 Throttle shaft bushing
51 Fuel transfer tube
52 Pump transfer tube assy.
53 Idle adjusting screw
54 Primary float & hinge assy.
55 Secondary float & hinge assy.
56 Fuel inlet needle & seat assy.
57 Fuel inlet fitting
58 Pump discharge nozzle
59 Primary main metering jet
60 Pump discharge nozzle check needle
61 Primary power valve assy.
62 Fuel transfer tube 0-ring seal
63 Choke rod seal
64 Secondary diaphragm check ball
65 Idle limiter cap
66 Secondary diaphragm housing cover
67 Pump cover assy.
68 Primary fuel bowl & plugs assy.
69 Secondary fuel bowl & plugs assy.
70 Primary metering body & plugs assy.
71 Secondary metering body
72 Secondary metering body plate
73 Choke housing & plugs assy.
74 Secondary diaphragm & link assy.
75 Pump diaphragm assy.
76 Secondary diaphragm link retainer
77 Float & hinge assy retainer
78 Pump lever retainer
79 Throttle stop screw spring
80 Idle adjusting needle spring
81 Pump diaphragm return spring
82 Primary float spring
83 Secondary float spring
84 Pump lever adjusting screw spring
85 Fuel inlet filter spring
86 Secondary diaphragm spring
87 Transmission kickdown lever spring
88 Choke shaft nut
89 Throttle modulator lock nut
90 Needle & seat adjusting nut
91 Pump lever adjusting screw nut
92 Choke thermostat lever spacer
93 Pump cam
94 Fast idle cam assy.
95 Choke rod
96 Throttle connecting rod
97 Choke housing shaft lock washer
98 Bracket screw lockwasher
99 Choke thermostat & cap assy.
100 Choke rod retainer
101 Throttle connecting rod retainer
102 Choke thermostat clamp
103 Throttle modulator bracket
104 Throttle modulator assy.
105 Primary fuel inlet baffle
106 Secondary fuel inlet baffle
107 Fuel inlet filter
108 Secondary diaphragm lever assy.
109 Pump operating lever
110 Secondary diaphragm housing
111 Idle solenoid assy.

MODEL 4500 Exploded View
Holley Typical View 42-1

NOTE: General view is useful for visualizing relationship of various parts in the carburetor. Specific details will vary with Part Numbers because each carburetor is made to fit a particular application.

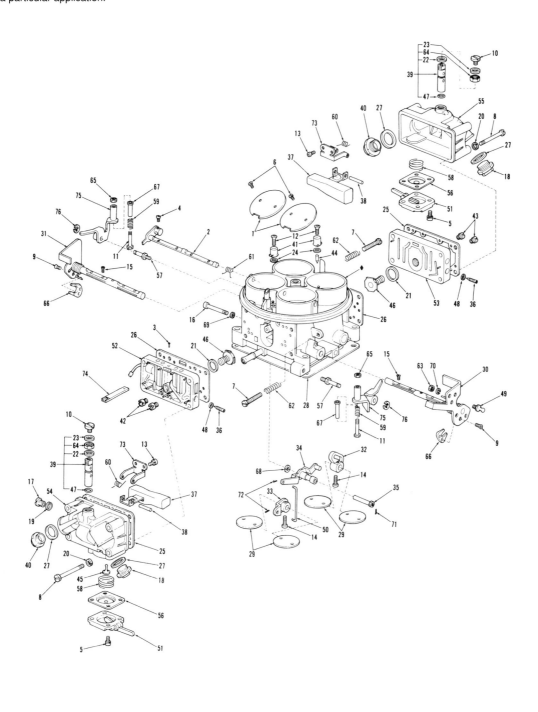

1. Choke plate
2. Choke shaft assy.
3. Fuel bowl vent baffle drive screw
4. Choke shaft swivel screw
5. Fuel pump cover screw & LW
6. Choke plate screw & LW
7. Throttle stop screw
8. Fuel bowl screw
9. Pump cam lock screw
10. Fuel valve seat lock screw
11. Pump operating adj. screw
12. Pump discharge nozzle screw
13. Float shaft bracket screw & LW
14. Throttle shaft screw
15. Throttle plate screw
16. Pivot screw
17. Fuel level check plug
18. Fuel inlet plug
19. Fuel level check plug gasket
20. Fuel bowl screw gasket
21. Power valve gasket
22. Fuel valve seat adj. nut gasket
23. Fuel valve seat lock screw gasket
24. Pump discharge nozzle gasket
25. Fuel bowl gasket
26. Metering body gasket—pri. & sec.
27. Fuel inlet fitting & plug gasket
28. Flange gasket
29. Throttle plate
30. Throttle shaft assy.—primary
31. Throttle shaft assy.—secondary
32. Primary throttle lever (internal)
33. Secondary throttle lever & bushing assy.
34. Intermediate throttle lever assy. (comp.)
35. Threaded guide bushing
36. Idle adjusting needle
37. Float & hinge assy.
38. Float shaft
39. Fuel valve & seat assy.
40. Fuel inlet fitting
41. Pump discharge nozzle
42. Main jet—primary
43. Main jet—secondary
44. Pump discharge needle valve
45. Pump check valve
46. Power valve assy.
47. Fuel valve seat O-ring seal
48. Idle adjusting needle seal
49. Throttle lever ball
50. Connecting link
51. Fuel pump cover assy.
52. Metering body & plugs assy.—primary
53. Metering body & plugs assy.—secondary
54. Fuel bowl—primary
55. Fuel bowl—secondary
56. Pump diaphragm assy.
57. Pump lever stud
58. Diaphragm return spring
59. Pump operating adj. screw spring
60. Float spring
61. Choke spring
62. Throttle stop screw spring
63. Throttle lever ball nut
64. Fuel valve seat adj. nut
65. Pump operating adj. nut
66. Pump cam
67. Pump operating lever screw sleeve
68. Pivot screw LW
69. Pivot screw washer
70. Throttle lever ball LW
71. Pivot screw cotter pin
72. Connecting link cotter pin
73. Float shaft retaining bracket
74. Fuel bowl vent baffle
75. Pump operating lever & guide assy.
76. Pump operating lever retainer

NOTE: The Model 2010 two-barrel is similar to the Model 4010 and 4011 because it is the primary half of these carburetors. Because of their similar construction they are discussed in one set of disassembly, repair and adjustment procedures. For demonstration purposes we used a Model 4010.

(1) Loosen two fuel-inlet fittings while carburetor is on the engine or on its "legs."

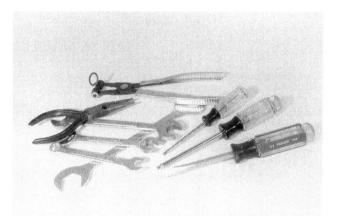

(2) These carburetors use T-20 and T-25 Torx head screws. Here are the tools you'll need: open-end wrenches, needle-nose pliers and screwdrivers—Phillips, flat-blade and T-20 and T-25 Torx drivers. Corbin-clamp pliers make it easy to remove wire hose clamps such as the one pictured. You'll also need feeler gauges.

(3) Use a T-25 Torx driver to remove screws holding airhorn to main body. There are 10 Torx screws on the 4010; the 4011 has eight; the 2010 has six.

(4) Remove center airhorn stud with a 7/16-inch wrench.

(5) Choke-actuating rod must be disconnected (C-clip or wire clip) at its lower end before airhorn can be removed.

(6) Lift airhorn straight up from main body to avoid float damage. Gasket is not usually sticky and airhorn should come loose easily. Gasket may be reusable, but we suggest using the new one from the kit.

(7) Remove float hinge pins and floats from both primary and secondary sides.

(8) When reinstalling floats and needle-and-seat assemblies, this approximate setting gets fuel level about right. Precise adjustment is done later.

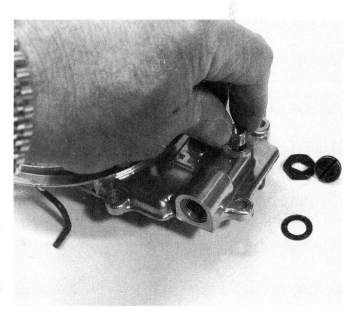

(9) Remove lock screws and adjusting nuts, then inlet needle-and seat assemblies from top. Install new ones from kit on reassembly. Use new washers and O-rings also. Big ones go on bottom.

(10) Remove two inlet fittings you previously loosened and remove filter screens.

(11) Use compressed air to blow out any foreign material from screens.

(12) Remove primary and secondary clusters. Each has a single holding screw. If a cluster is on a side that has an accelerator pump, that holding screw is hollow to allow fuel flow from the pump into the shooters. If there is no pump on that side, the holding screw is solid. Replace gaskets under each cluster and under the holding-screw head. Mark the primary cluster so you can put the clusters back in their original positions on reassembly. If the secondary side does not have an accelerator pump, there won't be any brass shooter restrictions in that cluster.

(13) Most metering restrictions for idle and main systems are in the clusters. Soak clusters in carburetor cleaner or lacquer thinner. Give each restriction a blast of carburetor cleaner, followed by a blast of compressed air.

(14) Removed primary cluster has hollow holding screw. Secondary cluster is still in place. Carburetors with secondary diaphragm do not have an accelerator pump on secondary side, so holding screw is solid and there's no ball and weight under screw. Invert the carburetor and catch the 1/8-inch-diameter steel rod (weight) and 3/16-inch steel ball. There are two weights and two balls on carbs with mechanical secondaries.

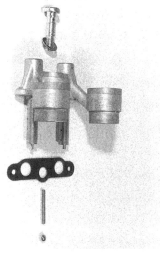

(15) Check this photo when reinstalling clusters. This assembly installs in a primary or secondary side that uses an accelerator pump. If the cluster serves a side without an accelerator pump, there are no shooters, the screw is solid and there is no ball or weight.

(16) Using a wide screwdriver to span the main-jet slot, remove main jets from bowls. Keep primary and secondary jets separate or note jet sizes so they can be reinstalled in the same locations on reassembly.

(17) T-20 Torx driver removes the four power-valve-cover screws. Lift off cover. Use a new gasket when you replace the cover.

(18) Remove power valve(s) with a 1-inch socket or you can use a box-end wrench if you are careful not to round off the flats.

177

(19) Shoot the power-valve passages with carburetor cleaner, then compressed air. Where there's no power valve, shoot cleaner and compressed air through main-jet passages.

(20) T-20 Torx driver removes the four pump-cover screws. Lift off pump cover, diaphragm and spring. Take out the rubber umbrella pump-inlet valve. Use a new diaphragm when you replace the cover.

(21) Spray pump inlet and outlet passages with cleaner and follow with compressed air.

(22) Replace rubber inlet valves with new ones in kit. Push stem through from bottom and pull from bowl side with needle-nose pliers until ball section pops through.

(23) Three T-20 Torx screws attach manual-choke assembly to main body. Remove assembly.

(24) Note how choke lever and rod are situated so you can reinstall actuating assembly correctly. To simplify getting the round extension into the spring-loaded lever nearest the main body, hold choke open and with actuating assembly in choke-open position. This assembly needs no cleaning unless it is coated with grease and grime. Carburetor cleaner won't affect the plastic parts.

(25) Electric choke has an index mark on the black bimetal housing. Note how this index relates to the graduations on the choke-housing casting. On reassembly, make the index location the same. If you failed to note the original position, set choke so it is just closed at room temperature.

(26) Remove three screws holding choke-cap retainer. Remove retainer and choke-bimetal cap.

(27) During reassembly, make sure bimetal loop engages lever tang. If you don't, bimetal can't rotate choke as temperature changes. Check whether loop is on tang by rotating choke cap first one way and then the other as you observe choke plate opening and closing.

(28) Remove three T-20 Torx screws attaching choke housing to main body. Note how choke relates to fast-idle cam so you can reassemble the same way. Remove choke assembly. Choke-actuating rod goes above fast-idle cam. Be careful not to trap fast-idle cam above the choke-actuating rod.

(29) Back side of choke assembly. Replace small cork gasket for vacuum passage (arrow) during reassembly.

(30) Now remove three T-20 Torx screws attaching secondary-diaphragm housing.

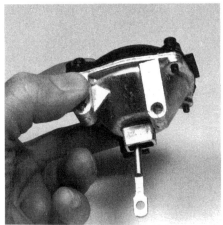

(31) Check diaphragm for leaks. Push stem all the way in and hold your thumb over vacuum passage. Release stem. If it moves, the diaphragm must be replaced.

(32) Disassemble diaphragm by removing four Phillips-head screws at top. Hold cover against the housing as you do this because diaphragm spring exerts outward force.

(33) Secondary-diaphragm unit disassembled. Replace diaphragm with new one provided in kit.

(34) Upon reassembly, make sure the ball check is inserted in hole as shown at top center (arrow).

(35) It is usually not necessary to proceed any further in disassembling the main body. Don't remove throttle plates unless it is absolutely essential. If you do remove them, be ultra careful. Retaining screws are staked as shown to prevent their loosening. Staking must be removed before removing the screws. Also, throttle shafts must be supported from the backside when restaking.

(36) It is not usually necessary to go any further with the airhorn. Upsets in the choke plate make disassembly difficult. If necessary, pry open slot in shaft with a screwdriver and pull out plate. Don't do it unless you have to.

(37) When airhorn is reassembled, set floats to approximate relationship as illustrated, step (8), page 175. Set final fuel level on the vehicle. Remove sight plug (arrow) from primary side. Place a shop cloth beneath the opening to catch spilled fuel. With engine at idle, loosen lock screw slightly with a screwdriver as shown. Turn adjustment nut clockwise to lower float and opposite to raise it. Set level just at the bottom of sight-plug opening. Flush bowl by revving engine slightly. You may have to make several attempts to get it right. Tighten lock screw. Replace plug and check for leaks. Now repeat the process on the secondary bowl. Once levels are set correctly, it is not usually necessary to reset them unless the carburetor is removed and disassembled.

(38) Idle speed is set as shown with engine fully warmed. Set speed to about 550 rpm with automatic transmissions in drive, or 650–700 rpm in neutral with manual transmissions.

(39) There are two idle-mixture screws, one on each bore of the primaries. Turn in (clockwise) to lean and opposite to richen. Start at either side and turn in until engine roughens slightly, then back out until idle smoothes. Stay a little on the lean side if the car must pass an emissions test. You can also set idle with a tachometer: lean mixture to drop engine speed 10 rpm from the max rpm attainable. After you've set one side, go to the other side and repeat the process. Now go back to the first side and do it again just to be sure. Recheck idle speed as described previously.

(40) Secondary "idle speed" adjustment just keeps the secondaries from sticking against their bores. Secondary throttle-plate adjustment is probably OK if secondaries don't stick. Otherwise, give throttle-adjustment screw one turn from where plates seat in bore. If you have a problem getting idle adjustability on the primary side, open the secondary plates a little more. This allows the primary plates to be closed slightly and cover the transfer ports a little more. If idle speed cannot be brought down to the desired rpm, close the secondary plates a little bit. But don't allow the secondary plates to seat in their bores because this will bind them and affect secondary opening, especially with vacuum secondaries.

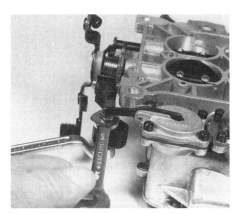

(41) Adjust screw on pump lever so lever contacts pump cam at idle. This ensures a pump shot the instant the throttle moves, avoiding sags or delays. On double-pumpers, adjust both primary and secondary levers.

(42) With throttle held wide open, there should be at least 0.010-inch clearance between the intermediate lever and the pump lever to avoid overstressing the diaphragm and bending the pump linkage. Check both primary and secondary for clearance.

(43) Choke qualify is the angle to which the choke is pulled immediately after the engine starts. The closer the choke plate is to vertical, the leaner the mixture. This position is determined by the stroke of a small piston inside cylinder shown. Stop is adjusted as shown. Turning adjustment screw clockwise shortens stroke and richens mixture. Turning counterclockwise leans mixture.

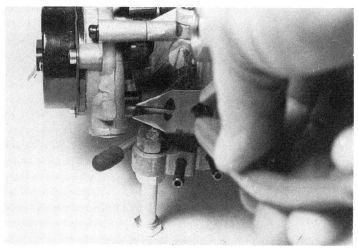

(44) Dechoke setting allows manifold to be cleared of fuel by opening choke approximately 1/3 of its travel when throttle is fully opened during cranking. Adjustment is seldom necessary, but can be done by bending tab as shown. Setting applies only to electric-automatic-choke versions with diaphragm-operated secondaries.

(45) Set fast idle to about 1700 rpm. Engine should be warm, choke held open and adjustment screw against top of cam or against top step with electric-automatic chokes.

2010-1/With airhorn removed, main jets and nozzle bar with shooters and booster venturis are accessible.

2010-2/ Accelerator-pump shooter and booster-venturi cluster removed.

2010-3/Underside of airhorn shows plastic float, choke, adjustable inlet valve, fuel inlet with screen and plug for alternate fuel inlet.

2010-4/Base of 2010 after disassembly of power valve and accelerator pump.

Holley 12-833 Street Performance mechanical pump for small-block Fords disassembled to reveal construction details. Also available for Chrysler and GM, these pumps feature 38-gph capacity at 5000 rpm with 4.5-psi back pressure. Large inlet/outlet ports (allows hose to 3/8-in. ID) in rotatable housings can be turned to simplify plumbing.

Holley 12-802 pump mounted on 12-809 insulator mounting bracket. Position near tank to push fuel at 9 psi to regulator 12-803 (left) at carburetor. 12-801 pump at right is shown with the standard rubber-lined clamp mount. Standard pressure regulator 12-804 is for fuel-pressure range from 1 to 4 psi. Pressure switch 12-810 senses engine oil pressure to shutoff pump when engine stops, whether ignition is ON or OFF.

Cutaway side view shows rotor (5) with sliding vanes (7), wear sleeve (8), wear plates (9) on top and bottom of pumping chamber, seal (10) and bronze bushing (11) for armature shaft.

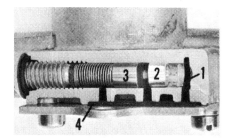

Pump passages provide output-pressure regulation. Cutaway of pressure-relief valve shows how this works. Fuel at higher-than-desired pressure from outlet side (1) enters hollow seat (2) to act against spring-loaded piston (3). Pressure pushes piston off seat so high-pressure fuel is routed back to inlet side of pump (4), thereby regulating pressure. Rotor (5) is offset in pumping chamber (6) formed by vanes (7) in wear sleeve (8).

Holley pump installation shows return line installed in location normally occupied by screw for pressure-relief valve. Hose contained a 0.060-in. restrictor. Arrow points to return line.

Two fuel-pressure gauges and two regulators on Pro Stock car. Fuel pressure can be observed during warmups in pits. Second regulator is hidden behind water hose; it is mount for electrical sender for fuel-pressure gauge inside cockpit.

Dual fuel-pressure regulators supply these Model 4500s on Pro Stock car.

WARNING: Avoid heavy bypass springs. A 19-psi bypass spring is being sold by another company for use in Holley's 12-802 max-pressure pump. Holley tests show that using this spring reduces pump life by 75%.

spring-loaded bypass valve which routes fuel from the outlet to the inlet side.

Power for the pump is supplied by a 12-volt permanent-magnet-type DC motor which runs dry. A seal on the motor shaft keeps fuel out of the motor. Bronze bearings are used at each end of the motor shaft.

Both street and performance versions of the pump are offered. Although most of the other components are the same, the 12-801 red low-pressure pump has a different bypass valve and a motor with a lower rating and a different winding. This pump is preset for 7 psi output. While it is designed for street and highway use without a fuel-pressure regulator it can be used with pressure regulator 12-804 or 12-500 (chrome fiinish) to regulate pressure from 1–4 psi. It is not designed for competition.

The 12-802 blue performance pump comes with a 12-803 fuel-pressure regulator that should be mounted near the carburetor/s. A spring operates against the regulator diaphragm to open the internal valve. Line pressure against the same diaphragm closes the regulator valve to maintain a preset pressure. Regulated output pressure can be externally adjusted from 4-1/2—9 psi by changing the spring-preload adjustment screw. A locknut maintains the setting. Pump and regulator inlets and outlets are tapped for 3/8-in. NPT.

VOLUMAX™ FUEL PUMPS

In 1991 Holley introduced two billet-style racing pumps using gerotor design and much larger motors than the other two Holley electric pumps. These precision machined pieces look like they "belong" in a racing machine. Output of the blue 12-705 pump is 160 gph; the white 12-706 pump puts out 250 gph.

Each pump has two 1/2-in. NPT inlet ports with screens, a 1/2-in. NPT outlet port and a 3/8-in. NPT return port. Both pumps include mounting brackets.

214

These pumps are not designed for continuous use. They are strictly race-car hardware.

Each pump has an internal rotor that is driven by the electric motor. An external rotor is turned by the internal rotor. Fuel enters at one side where the spaces between the rotors are increasing. Fuel is carried around to the other side where the spaces are decreasing. The lobes of the internal rotor move into the spaces and force the fuel out.

The matching 12-704 aluminum-finish billet-style regulator reduces the pumps' 15-psi output pressure to 4-1/2—9 psi, as desired. The regulator has one 1/2-in. NPT inlet port and two 1/2-in. NPT outlet ports. A mounting bracket is supplied with the regulator.

These pumps should be plumbed with a filter in each inlet line. Holley recommends the Fram HPG-1 or the AC-Delco T-95 filters in each inlet line to the Volumax pumps. These filters should be carefully maintained so they don't plug-up with dirt and rust and become restrictive.

MARINE PUMP

A third type of pump is offered for stock marine applications. This is the 50-gph Volumax billet-style marine pump, part 712-703. This gerotor unit also has a fuel/fume vent tube to redirect gasoline and/or vapors back into the carburetor to help prevent explosive build-up in the engine compartment. Output is set at 6 psi, so no external regulator is required. The motor is the same as used on the smaller Holley electric pumps. It should be used with a non-restrictive filter on the inlet.

Disassembled Volumax pump shows gerotor, inlet and outlet ports to gerotor in pump cover and check valve. This pump was photographed before all of the parts were black-anodized so the details would show up better.

 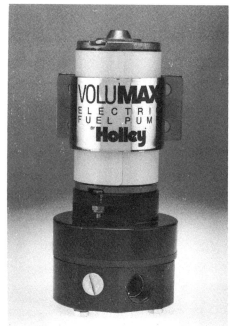

Holley Volumax billet-construction 160-gph (12-705) (left) and 250-gph (12-706) fuel pumps .

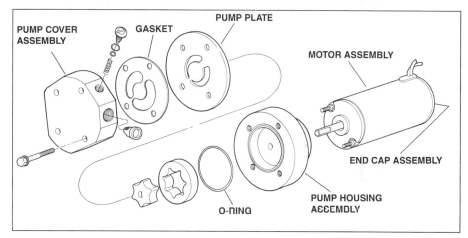

Exploded view of Volumax pump (typical for both pumps) shows six-lobe internal rotor that is turned by the motor, driving the external rotor. Pump-plate openings are the inlet and outlet to the rotors.

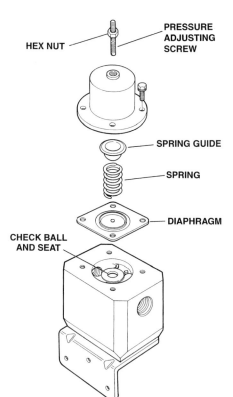

The 12-704 fuel pressure regulator (left) has one 1/2-in. NPT intlet port and two 1/2-in. outlet ports for use with a single four-barrel. The 12-707 regulator has four AN-6 outlets and one AN-8 inlet for use with two four-barrels.

Exploded view of Volumax regulator shows how spring force is applied to atmospheric side of diaphragm. Wet side of the diaphragm has a pin that bears against the check ball. This is an early version of the 12-704. It was subsequently changed from billet to die-cast construction.

INSTALLING AN ELECTRIC FUEL PUMP

Wiring—First, check that the circuit you'll be using for the electric pump provides 13.5 volts—the nominal regulated voltage of 12-volt systems when the engine is running. Pump performance drops dramatically when voltage drops. At 13.5 volts the 12-801 and 12-802 pumps draw approximately 3 amperes, so include a 7.5-ampere fuse in the circuit that powers the pump. The 12-705 and 12-706 pumps require 7.5- and 15-ampere fuses, respectively.

Include a safety switch in the circuit so the pump won't work unless there is oil pressure. Holley's 12-810 switch can be mounted on a 1/8-in.

pipe tee. Put the stock pressure switch for the idiot light on the other side. The fuel pump will shut off if the engine stalls with the ignition ON.

Using the switch to turn the engine with the starter energizes the pump from the starter solenoid. Once the engine is running, the switch provides voltage to the pump so long as there is oil pressure to keep the switch turned on. A schematic showing the switch wiring circuit is included at right.

Mounting—Place the pump in as cool an area as possible; keep it away from any exhaust-system components. Any fuel pump mounted to the chassis will transmit some noise into the car's body structure. While this is no problem on a race car with open exhaust, it can be annoying on a dual-purpose car driven on the street.

Reduce chassis-transmitted noise from a 12-801 or 12-802 pump by using Holley's 12-809 insulator mounting bracket. And for further noise reduction, mount this bracket on four rubber-insulated studs mounted on a chassis member or onto a stiff or reinforced body section.

Never mount the pump onto a large flat sheet-metal surface because this amplifies pump noise. Always mount the pump near the fuel tank (usually at the rear of the car) and as low as possible so suction height is kept to a minimum. It is best if the pump is gravity-fed from the tank.

Creating a vacuum (low absolute pressure) on the end of a fuel line, especially a long one, tends to allow the fuel to flash into vapor, which is very hard to pump. That's why professional racers replace the engine-mounted mechanical pump with an electric pump near the fuel tank; it's positive insurance against vapor lock.

When fuel is pumped forward at high pressure, there's another advantage: this offsets pressure losses caused by line loss (friction) and by acceleration (g forces). Pumping the fuel forward at high pressure and reducing pressure prior to feeding it to the carburetor/s ensures there will always be adequate pressure available at the carburetor/s. This greatly reduces the tendency to vapor lock.

Fuel Lines—Keep the pump-inlet line short and large. If a filter is used between the fuel tank and the pump inlet, it must be a screen type or a Fram HPG-1 or AC-Delco T-95. There must not be any pressure drop at this point. That could cause fuel vaporization and consequent pump cavitation. Position the pump so its inlet is slightly below and behind the tank so the fuel level creates a *head* or positive pressure at the inlet. Acceleration forces will tend to keep fuel at the inlet.

NOTE: No electric pump should ever be allowed to run dry.

Stock-size (5/16 or 3/8-in.) steel fuel lines can be used on a street machine, but they compromise pump performance. Small lines, sharp bends or kinks and right-angle fittings will cause a pressure drop.

Keep the lines away from exhaust-system components to avoid excessive heat. Make sure no part of the body can deflect exhaust from open headers

back onto the line. If a line passes near the exhaust system and there is no other place to route it, insulate the fuel line and build a heat shield to go between the heat source and the fuel line. Clamp the line against the chassis or body structure with rubber-lined aircraft clamps.

Pressure Regulator—Mount the pressure regulator as close to the carburetor/s as possible. Lines between the regulator and carburetor/s can be 3/8-in. ID or AN-6. There are two outlets on the 12-803, 12-804 and 12-704 regulators. If the carburetor has bowls with individual inlets, connect each bowl to an outlet. Where two carburetors are used, connect each carburetor to an outlet. Some racers prefer using two regulators, one at each carburetor. Or use regulator 12-707 that has four AN-6 outlet ports.

Pressure at the carburetor must be set with the engine idling so there will be some flow to allow the regulator to function. Use a fuel-pressure gauge at the carburetor and adjust the regulator to provide 6–7 psi output (factory-set when the 12-803 blue regulator is tested).

Cool Can—When running the car at high ambient temperatures, use a cool

can just ahead of the regulator (upstream) on the high-pressure side so fuel won't tend to flash into vapor when it is changed to a lower pressure by the regulator.

Fuel Return—Some racers install a separate return (bypass) from the 12-802 pump outlet or pump-relief valve to the tank. Then the pump does not continually pump the same fuel in a loop from the pump bypass back to the inlet. An external bypass (return) allows fresh fuel to appear at the pump inlet, even in low-demand situations such as idling.

The bypass for 12-801 or 12-802 pumps should be through a 1/16-in. (or 0.060) restriction. Don't use any restrictor larger than this! Even with this bypass arrangement, capacity of the high-performance pump is 550 lbs per hr of fuel at 9 psi.

The Volumax 160- and 250-gph pumps have a 3/8-in. NPT return port

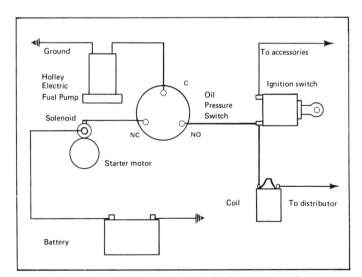

Typical wiring diagram for electric fuel pump. Holley's pressure switch 12-810 ensures pump won't operate unless engine is running or ignition switch is energizing starter solenoid. Starter circuit from ignition switch not shown in drawing.

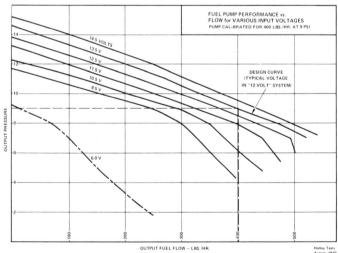

Note how 12-802 pump performance drops off as voltage drops. This emphasizes the importance of keeping the battery charged. Pumps are designed to operate at 13.5 volts—nominal regulated voltage of 12-volt systems. At 13.5 volts the pump draws approximately 4 amperes.

Rust accumulation in the bottom of electric pump is not unusual. It's a good idea to drain the tank and the complete fuel system before storage. Disconnect the pick up line to the pump and squirt in some motor oil so the vanes won't freeze in the rotor. Take the bottom off the pump occasionally and clean out any accumulated dirt and rust.

(RET) to simplify this plumbing. Do not use a restrictor in the return line. Always use a return line with these pumps.

ELECTRIC FUEL PUMP SERVICE KITS

Service kits for the 12-801 and 12-802 pumps allow repairing leaks, brushes, pump rotor and vanes and the check valve and filter screen.

Kit 12-808, the Lower Housing Service Kit) includes a new pump housing with the seal assembly positioned so it will ride on a previously unused portion of the shaft. The rotor, pump vanes, motor and bottom cover are removed from the old pump and installed in the new housing.

Kits 12-805 and 12-806 include a pump check-valve assembly and an inlet screen. Rotor Service Kit 12-811 includes a rotor and vanes. Armature Cap & Brush Kit 12-855 is used to fix pumps with worn brushes.

Diaphragm Repair Kit 12-807 is used to repair 12-803 and 12-804 pressure regulators.

WATER IS THE ENEMY!

Holley's electric pumps are extremely reliable. Yet Holley Technical Representatives fix several at every drag race where Holley is represented. There is always a rash of failures at the first races of the season.

Inevitably, failures are caused by water in the fuel. Where does the water come from? Humid conditions

are common in most of the U. S. and Canada, especially during the summer months. Water condenses on the tank surfaces and falls into the fuel, settling in the bottom of the tank and in the pump. Sometimes, water is pumped into your tank with the gasoline.

Because the pump is usually the lowest point in the fuel system, water settles in the pump and sits there, causing corrosive damage. We see pumps corroded so badly that the rotor won't spin. Vanes are stuck in the rotor and the pressure-relief valve is stuck tight.

Although these problems are easily fixed with Holley's service parts kits, it takes time. And it's aggravating to miss out on the racing.

FIRE—A HAZARD YOU CAN MINIMIZE
by Howard Fisher

Carburetors and fuel-supply system components carry or contain gasoline—a very flammable liquid. Careful installation and observation of what is happening when fuel pressure is applied to the carburetor (watching for flooding, etc.) greatly reduces the possibility of fire.

Workplace—Where a car or a carburetor is being worked on in a closed area containing a flame—such as the pilot light or burner of a water heater or furnace—**fire danger is extreme!** One spill can generate sufficient vapor to be carried across the floor to the flame. Then the trouble begins in the form of an explosion or a fire or both. Remember—any carburetor removed from a car contains gasoline. Drain it before carrying the carburetor inside to work on it. This is especially true when you work in an area with any kind of open flame.

Air Cleaner—75% of all auto fires directly result from leaving off the air cleaner. Fuel spews out of the carburetor onto the—manifold—or standoff collects on the underside of the hood—then a backfire ignites the fuel.

Fire experts point out that the air cleaner prevents a backfire from igniting any stray fuel. The air cleaner itself does not support combustion very well. It's interesting to note that flame arresters are required on the carburetors of all marine inboard gasoline engines. If a fire starts in the air cleaner, a lot of smoke can be expected, but not much burns if the element is the usual paper type. Nearly all auto fires not caused by leaving the air cleaner off are due to a bad fuel line or connection between the fuel pump and carburetor.

What To Do—Immediately call the fire department when a fire breaks out. Don't wait until you've used up your extinguisher and the fire is not out. Get the fire department on the way just in case.

If you have a fire under the hood, don't throw the hood open because hot gases and flames will rush out. Open the hood just a crack and shoot the extinguishing media in. Better yet, shoot it in from under the engine.

Check everything before restarting the engine or you could start a worse fire—just when your extinguisher is all used up. Check all ignition wires, battery cables and the fuel lines to see whether they have burned or melted. Check fuel-line connections to make sure they have not loosened.

Fire Extinguishers—If a fire extinguisher is carried in the car, the fire can usually be put out quickly and with little damage. A 2-pound ABC dry-chemical extinguisher covers all three classes of fire found in cars. A is for upholstery and the interior, which are common burnables, B refers to gasoline/oil fires, C is an electrical fire.

Dry Chemical—Volume-for-volume, powder has much better extinguishing capabilities than carbon dioxide (CO_2) or Halon FE-1301 (Bromotrifluoromethane). Also, powder retards reignition of the fire at hot spots and wiring.

Drawbacks of the dry-chemical extinguisher are that the powder goes everywhere, leaving a mess to be cleaned up afterward. And, the powder can get in the engine, especially if there is no air cleaner. If a lot of powder has to be directed into the carburetor air inlet, some may get into one or more cylinders through an open intake valve. When you attempt to restart, a piston will compress this into a cake which may prevent the engine from turning. Even if you can crank the engine, the powder is abrasive. So, if very much gets in the engine, pull the cylinder heads and clean out the powder before running the engine.

Carbon Dioxide—CO_2 extinguishers are preferred by many because they cannot damage the engine and there is no after-mess. This is their greatest plus. But, because CO_2 fights fire by displacing oxygen, a much larger extinguisher (than dry chemical) is required to match the power of a dry-chemical unit. Fire experts we talked with recommended a 50-pound CO_2 unit! In open areas, especially when the wind is blowing, CO_2 dissipates very quickly and sometimes will not put out a burning fuel line or wire.

Halon—An alternative to CO_2 is Halon FE-1301. This is the only approved extinguishing agent. Chemical products formed when Halon FE-1301 is exposed to flame interrupt the reactive process essential to fire. A concentration of only 4% (by volume) of Halon FE-1301 extinguishes the flames of most common fuels.

While the makers of systems using this extinguishing agent claim that a concentration of 20% can be breathed safely with no ill effects, material from the National Fire Protection Association disagrees. The NFPA says, "Halon 1301 vapor has low toxicity. However, decomposition products (as in a fire) can be hazardous. When using these extinguishers in unventilated places such as small rooms, closets, motor vehicles or other confined spaces, avoid breathing gases produced by thermal decomposition of Halon 1301."

Halon dissipates afterward with no mess or harm to the engine or other components. Pound-for-pound, Halon has three times the extinguishing power of CO_2. Even so, it dissipates rapidly in open areas or where the wind is blowing.

A 5-pound Halon extinguisher is an excellent item to carry in your car or have handy in your garage. If you have a CO_2 or Halon and a dry-chemical extinguisher on hand, always use the CO_2 or Halon first.

Many racecars are equipped with on-board Halon FE-1301 extinguisher systems. The added safety margin this type of extinguishing system gives the driver who may have to exit from a flaming vehicle offsets their seemingly high first cost.

They pipe pressurized Halon into the driver's cockpit as low and far forward as possible. To avoid frostbite, keep the cockpit nozzle 18-inches away from any part of the driver. This location reduces the possibility of air flow sucking the Halon out of a low-pressure. area. Additionally, nozzles direct Halon over, under and around the engine.

Index

HANDBOOKS

Auto Electrical Handbook: 0-89586-238-7
Auto Upholstery & Interiors: 1-55788-265-7
Brake Handbook: 0-89586-232-8
Car Builder's Handbook: 1-55788-278-9
Street Rodder's Handbook: 0-89586-369-3
Turbo Hydra-matic 350 Handbook: 0-89586-051-1
Welder's Handbook: 1-55788-264-9

BODYWORK & PAINTING

Automotive Detailing: 1-55788-288-6
Automotive Paint Handbook: 1-55788-291-6
Fiberglass & Composite Materials: 1-55788-239-8
Metal Fabricator's Handbook: 0-89586-870-9
Paint & Body Handbook: 1-55788-082-4
Sheet Metal Handbook: 0-89586-757-5

INDUCTION

Holley 4150: 0-89586-047-3
Holley Carburetors, Manifolds & Fuel Injection: 1-55788-052-2
Rochester Carburetors: 0-89586-301-4
Turbochargers: 0-89586-135-6
Weber Carburetors: 0-89586-377-4

PERFORMANCE

Aerodynamics For Racing & Performance Cars: 1-55788-267-3
Baja Bugs & Buggies: 0-89586-186-0
Big-Block Chevy Performance: 1-55788-216-9
Big Block Mopar Performance: 1-55788-302-5
Bracket Racing: 1-55788-266-5
Brake Systems: 1-55788-281-9
Camaro Performance: 1-55788-057-3
Chassis Engineering: 1-55788-055-7
Chevrolet Power: 1-55788-087-5
Ford Windsor Small-Block Performance: 1-55788-323-8
Honda/Acura Performance: 1-55788-324-6
High Performance Hardware: 1-55788-304-1
How to Build Tri-Five Chevy Trucks ('55-'57): 1-55788-285-1
How to Hot Rod Big-Block Chevys:0-912656-04-2
How to Hot Rod Small-Block Chevys:0-912656-06-9
How to Hot Rod Small-Block Mopar Engines: 0-89586-479-7
How to Hot Rod VW Engines:0-912656-03-4
How to Make Your Car Handle:0-912656-46-8
John Lingenfelter: Modifying Small-Block Chevy: 1-55788-238-X
Mustang 5.0 Projects: 1-55788-275-4

Mustang Performance ('79–'93): 1-55788-193-6
Mustang Performance 2 ('79–'93): 1-55788-202-9
1001 High Performance Tech Tips: 1-55788-199-5
Performance Ignition Systems: 1-55788-306-8
Performance Wheels & Tires: 1-55788-286-X
Race Car Engineering & Mechanics: 1-55788-064-6
Small-Block Chevy Performance: 1-55788-253-3

ENGINE REBUILDING

Engine Builder's Handbook: 1-55788-245-2
Rebuild Air-Cooled VW Engines: 0-89586-225-5
Rebuild Big-Block Chevy Engines: 0-89586-175-5
Rebuild Big-Block Ford Engines: 0-89586-070-8
Rebuild Big-Block Mopar Engines: 1-55788-190-1
Rebuild Ford V-8 Engines: 0-89586-036-8
Rebuild Small-Block Chevy Engines: 1-55788-029-8
Rebuild Small-Block Ford Engines:0-912656-89-1
Rebuild Small-Block Mopar Engines: 0-89586-128-3

RESTORATION, MAINTENANCE, REPAIR

Camaro Owner's Handbook ('67–'81): 1-55788-301-7
Camaro Restoration Handbook ('67–'81): 0-89586-375-8
Classic Car Restorer's Handbook: 1-55788-194-4
Corvette Weekend Projects ('68–'82): 1-55788-218-5
Mustang Restoration Handbook('64 1/2–'70): 0-89586-402-9
Mustang Weekend Projects ('64–'67): 1-55788-230-4
Mustang Weekend Projects 2 ('68–'70): 1-55788-256-8
Tri-Five Chevy Owner's ('55–'57): 1-55788-285-1

GENERAL REFERENCE

Auto Math:1-55788-020-4
Fabulous Funny Cars: 1-55788-069-7
Guide to GM Muscle Cars: 1-55788-003-4
Stock Cars!: 1-55788-308-4

MARINE

Big-Block Chevy Marine Performance: 1-55788-297-5

HPBOOKS ARE AVAILABLE AT BOOK AND SPECIALTY RETAILERS OR TO ORDER CALL: 1-800-788-6262, ext. 1

HPBooks
A division of Penguin Putnam Inc.
375 Hudson Street
New York, NY 10014